methods of analysis for PETROCHEMICALS

edited by

E. R. LITTMANN, PH.D.

Enjay Company, Inc.

1958

CHEMICAL PUBLISHING CO., INC.

212 Fifth Avenue New York, N.Y.

Methods of Analysis for Petrochemicals

ISBN: 978-0-8206-0021-5

Chemical Publishing Company:
www.chemical-publishing.com
www.chemicalpublishing.net

First edition:

© **Chemical Publishing Company, Inc.** – New York, 1958

Second Impression:

Chemical Publishing Company, Inc. - 2011

Printed in the United States of America

FOREWORD

The rapid increase in the use of petrochemicals as raw materials by the chemical industry has created a unique problem, since many methods of analysis for these products are generally not available. In many instances little or no effort has been made to standardize such methods as do exist while in others the purchaser hardly knows where to start in developing an adequate procedure of his own.

The term *petrochemicals* has in recent years been identified with a very large number of organic chemicals manufactured in the United States. A more accurate definition would, perhaps, include only the primary and secondary products derived from petroleum and used in subsequent manufacturing operations or synthesis steps. This second definition would cover such products as gaseous or liquid olefins and diolefins, alcohols, aromatics, e.g., benzene, toluene, xylene, certain synthetic alcohols, ketones, etc.

Many of the methods of the American Society for Testing Materials have been applied for quality evaluation of petrochemicals. In some cases, these methods could be used without modification, whereas in others, changes were necessary to adapt the methods to special products or to meet customers' specifications. Where methods were nonexistant, new ones had to be devised.

The methods set forth in this book have been assembled from those published by the American Society for Testing Materials and those developed in the laboratories of the Enjay Company, Inc. These methods are the ones used in determining the specification limits controlling the quality of the products sold by the Enjay Company, Inc., and are presented in an effort to provide the chemical industry with dependable standard procedures for the analysis of petrochemicals.

The author, in his capacity as editor and compiler is deeply appreciative of the permission granted by the American Society for Testing Materials to reproduce a number of their standard procedures and to his associates for their invaluable assistance in this work. Any comments regarding this book or suggestions for its improvement by the chemists and engineers who will have occasion to consult it will be greatly appreciated.

TABLE OF CONTENTS

DISTILLATION OF NAPHTHA, GASOLINE, KEROSENE, JET FUEL, AND SIMILAR PRODUCTS

The method of test recommended for the distillation of light petroleum liquids is exactly the same as that described in the current version of ASTM D 86, subject to the qualifications outlined in the next section.

D 86 covers a wide range of products. For best results, certain modifications should be made, depending on the nature of the sample, as follows:

KEROSENES

The specifications for the Engler flask permit a volume of the bulb in the range of 125 to 145 ml. To prevent trouble from foaming, the volume should be in the range of 135 to 145 when testing kerosenes. A special distillation flask, with a volume in this range should be reserved for kerosenes only.

Special Solvents

Whenever the sample under test contains peroxides or other oxycompounds, there is a possibility that the flask might explode toward the end of the test. If the presence of peroxides is suspected, make a preliminary determination of peroxide number, following L.I.C. No. 300.10B, included in this book. If the observed number is above 10, add 1 g of pyrogallol to the sample in the distillation flask before starting the test. In addition, put some sort of shield made of safety glass or clear plastic around the heating assembly.

The current version of Federal test method 1001, as described in *Federal Test Methods Std. 791* (formerly VV–L–791e), is exactly the same as ASTM D 86. The procedure outlined here should, therefore, be followed for all products intended for sale to the Federal government, unless otherwise dictated by the applicable specification.

Whatever the type of sample under test, the receiving graduate must be partly immersed in a bath of cold water. On humid days, the outside of the bath will be coated with a film of condensed moisture, making difficult the reading of the volume in the graduate. This difficulty can be overcome by wiping the outside of the battery jar with a cloth soaked in either alcohol or a mixture of alcohol and glycerin.

It has been noticed repeatedly that erratic results may be obtained when a new flask is used for the first few times. There seems to be a glaze on the surface of the glass which inhibits the formation of small bubbles of vapor and, during the early stages, there may be sudden spurts of gas. Trouble from this source can be eliminated by either of the following:

(a) Pour a few grams of carborundum or fine sand into the flask and agitate by hand shaking or a stream of compressed air. The agitation should be just sufficient to break up the glaze, but not strong enough to cause a visible etching or scratching.

(b) Fix a wad of fine steel wool to the end of a glass rod, insert the end into the flask, and rub the area immediately under the neck of the flask.

Several suggestions have been made for removing the carbon deposit inside the flask after use, as follows:

(a) Clean the flask with hot caustic solution (1).

(b) Rub the carbon deposit with the steel wood wad described under (6) and repeat any normal cleaning operation.

(c) Make a coil of copper tubing having at least two complete circles of a diameter slightly larger than that of the flask. Mount the flask slightly above, but not touching, the coils. Attach one end of the tubing to a source of compressed air and arrange the other end so that the air will be directed toward the bottom of the flask. Turn on the air, heat the *tubing* by a gas burner, and direct the stream of hot air toward the carbon deposits. Do not heat the flask, since this may result in a distortion of the glass.

ASTM DESIGNATION: D 86–56

Scope

1. This method of test is intended for use in the distillation of gasoline, naphtha, kerosene, and similar petroleum products.

Apparatus

2. The apparatus shall consist of the following:

(*a*) *Flask*—A standard 100-ml Engler distillation flask, as shown in Figure 1.

The position of the vapor tube shall be 9 ± 0.3 cm (3.55 ± 0.12 in.) above the surface of the liquid when the flask contains its charge of 100 ml. The tube shall be approximately in the middle of the neck and set at an angle of 75 ± 3° with the vertical.

(*b*) *Condenser*—A condenser (Figure 2 shows one of the approved types $\frac{9}{16}$ in. (14.29 mm) in outside diameter with 0.031 to 0.036-in. wall thickness, made of seamless brass tubing, 22 in. (55.88 cm) in length. It shall be set so that approximately 15.5 in. (39.4 cm) of the tube will be in contact with the cooling medium contained in the condenser jacket, subsequently described, with about 2 in. outside the cooling bath at the upper end, and 4.5 in. outside the cooling bath at the lower end. The length of tube projecting at the upper end shall be straight and shall be set at an angle of 75° with the vertical. The section of the tube inside the cooling bath may be either straight or bent in any suitable continuous, smooth curve. The average gradient shall be 0.26 in. per linear inch of condenser tube (sine of angle of 15°) and no section of the immersed portion of the condenser tube shall have a gradient less than 0.24 in. nor more than 0.28 in. per linear inch of tube. The projecting lower portion of the condenser tube shall be curved downward for a length of 3 in. (7.62 cm) and slightly backward so as to insure contact with the wall of the receiving graduate at a point approximately 1 to $1\frac{1}{4}$ in. (2.54 to 3.18 cm) below the top of the graduate when it is in position to receive the distillate. The lower end of the condenser tube shall be cut off at an acute angle. The capacity of the cooling bath shall be not less than 340 cu in. (5.55 liters) of cooling medium. The arrangement of the tube in the cooling bath shall be such that its center line shall be not less than 1¼ in. below the plane of the top of the bath at its point of entrance and not less than ¾ in. above the floor of the bath at its exit. Clearances

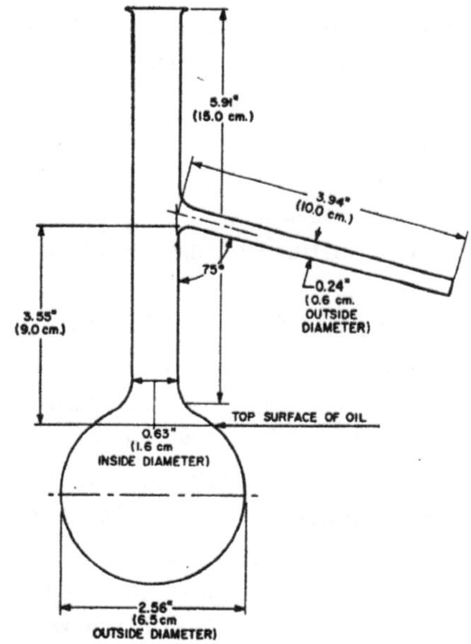

FIGURE 1. Dimensions of a 100-ml Engler Flask

between the condenser tube and the walls of the bath shall be at least ½ in. except for the sections adjacent to the points of entrance and exit. Multiple installations are permissible, provided they conform to the dimensional requirements and the capacity of the bath is not less than 340 cu in. per tube.

(*c*) *Shield*—Two types of shields can be used, as follows:

Type 1 Shield—A shield (Figure 2) 19 in. high, 11 in. long, and 8 in. wide, made of sheet metal approximately No. 22 gage, with a door on one narrow side, two openings 1 in. in diameter equally spaced in each of the two narrow sides, and with a slot cut in one side for the vapor tube. The centers of these four openings shall be 8½ in. below the top of the shield. There shall be three ½-in. holes in each of the four sides with their centers 1 in. above the base of the shield.

Type 2 Shield—A shield 17½ in. high, 8 in. long, and 8 in. wide, made of sheet metal of approximately No. 22 gage, with window on front side. Open bottom of shield spaced approximately 2 in.

from base of unit. Rear of shield to have eliptical hole for vapor tube. A flask adjusting knob located on the front of the shield for adjusting the flask support. Also heat-adjusting indicating dial to provide stepless heat control when electric heater is used. All mechanism built into bottom portion of shield.

(*d*) *Flask Support and Ceramic or Asbestos Boards*—A ring support of the ordinary laboratory type, 4 in. or larger in diameter, supported on a stand inside the shield, or a flask support platform that is adjustable from the outside of the shield may be used. Two ceramic or hard asbestos boards, the first having a hole $1\frac{1}{4}$-in.[1] in diameter in the center and the thickness of the board at the $1\frac{1}{4}$-in. diameter opening shall be $\frac{1}{8}$ to $\frac{1}{4}$-in.; the second board to fit tightly inside the shield with an opening 3 to 4 in. in diameter, concentric with the flask support. These

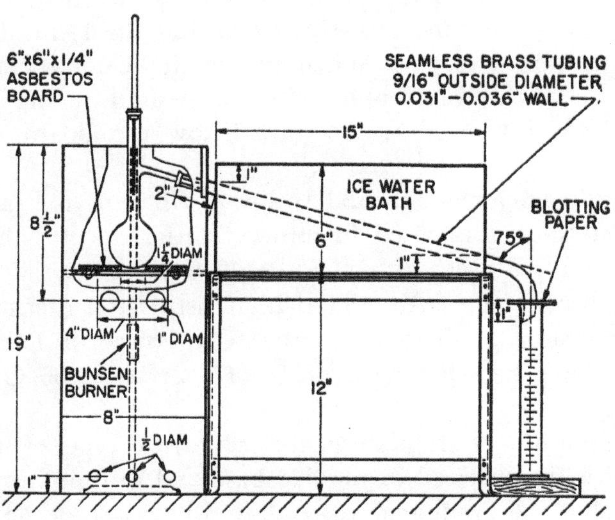

FIGURE 2. Apparatus for Distillation Test

shall be arranged as follows: The second board shall be placed on the support and the first or smaller board on top, so that it may be moved in accordance with the directions for placing

[1] When distilling petroleum products having an end point above 470°F (243.3°C), the hole in the asbestos or ceramic board shall be $1\frac{1}{2}$ in. (3.81 cm) in diameter.

the distillation flask. Direct heat shall be applied to the flask only through the 1¼-in. diameter opening in the first board.

(*e*) *Gas Burner*—A burner so constructed that sufficient heat can be obtained to distill the product at the uniform rate specified in Section 4. The flame shall never be so large that it spreads over a circle of diameter greater than 3½ in. (8.89 cm) on the under surface of the asbestos board. A sensitive regulating valve and gas-pressure governor are desirable adjuncts, as they give complete control of heating.

(*f*) *Electric Heater*—An electric heater may be used instead of a gas burner and shall be capable of bringing over the first drop within the time specified in Section 4(*a*) when started cold, and of continuing the distillation at a uniform rate. (A heater unit of low heat capacity, adjustable from 0 to 750 w, has been found satisfactory.) The electric heater shall be fitted with a ceramic or hard asbestos board top having a hole 1¼ in.[1] in diameter in the center, and the thickness of the board at the 1¼-in. diameter opening shall be ⅛ to ¼ in. When an electric heater is employed, the portion of the shield above the board shall be the same as with the gas burner, but the part below the board may be omitted.

(*g*) *Thermometer*[2]—An ASTM low-distillation thermometer, total immersion, graduated in either Centigrade or Fahrenheit degrees, having a range of −2° to +300°C or 30° to 580°F.

(*h*) *Thermometer*[2]—An ASTM high-distillation thermometer, total immersion, graduated in either Centigrade or Fahrenheit degrees as specified, having a range of −2° to +400°C or 30° to 760°F.

(*i*) *Graduate*[3]—A graduate of the cylindrical type, of uniform diameter, with a pressed or molded base and a lipped top. The cylinder shall be graduated to contain 100 ml, and the graduated portion shall be not less than 7 in. (17.78 cm) nor more than 8 in. (20.32 cm) in length. It shall be graduated in single milliliters and each fifth mark shall be distinguished by a longer line. It shall be numbered from the bottom up at intervals of 10 ml. The over-all height of the graduate shall be not less

[2] Whenever thermometers complying with ASTM requirements are not available, thermometers complying with the requirements of the Institute of Petroleum IP 5C (low distillation) and IP 6C (high distillation) may be used, provided calibration corrections are made in temperature ranges where IP requirements for scale accuracy are less stringent than those of ASTM.

[3] Cylinders lacking the bottom 1-ml line will not be rejected.

than 9¾ in. (24.8 cm), nor more than 10¼ in. (26.0 cm). The graduations shall not be in error by more than 1 ml at any point on the scale.

Preparation of Apparatus

3. (*a*) The condenser bath shall be filled with cracked ice,[4] and enough water added to cover the condenser tube. The temperature shall be maintained at 32° to 34°F (0° to 1.1°C) for gasolines higher than 9.5 lb Reid vapor pressure and 32° to 40°F (0° to 4.5°C) for other products.

(*b*) The condenser tube shall be swabbed to remove any liquid remaining from the previous test. A piece of soft, lint-free cloth attached to a cord or copper wire may be used for this purpose.

(*c*) A 100-ml sample shall be measured in the 100-ml graduated cylinder at 55° to 65°F (12.8° to 18.3°C) and transferred directly to the distillation flask which has been cooled at 55° to 65°F (12.8° to 18.3°C) for gasolines higher than 9.5 lb Reid vapor pressure. The flask may be at room temperature for other products. None of the liquid shall be permitted to flow into the vapor tube.

(*d*) The thermometer[5] provided with a cork shall be cooled to 55° to 65°F and fitted tightly into the flask so that it will be in the middle of the neck and the lower end of the capillary tube is on a level with the inside of the bottom of the vapor-outlet tube at its junction with the neck of the flask. The thermometer shall be approximately at room temperature when placed in the flask.

(*e*) The charged flask shall be placed in the 1¼-in. (3.18-cm) opening in the 6 by 6-in. (15.24 by 15.24-cm) asbestos board with the vapor-outlet tube inserted into the condenser tube. A tight connection may be made by means of a cork through which the vapor tube passes. The position of the flask shall be so adjusted that the vapor tube extends into the condenser tube not less than 1 in. (2.54 cm) nor more than 2 in. (5.08 cm).

(*f*) The graduated cylinder used in measuring the charge shall

[4] Any other convenient cooling medium may be used.

[5] Either the ASTM high- or low-distillation thermometer shall be used, depending on the initial boiling point and the end point of the product. Use the low-distillation type for all liquids except those found to have both the initial boiling points higher than 212°F (100°C) and final boiling points higher than 482°F (250°C) as determined using that thermometer.

be placed, without drying, at the outlet of the condenser tube in such a position that the condenser tube shall extend into the graduate at least 1 in. (2.54 cm), but not below the 100-ml mark. Unless the room temperature is between 55° and 65°F (12.8° and 18.3°C) the receiving graduate shall be immersed up to the 100-ml mark in a transparent bath maintained between these temperatures. The top of the graduate shall be covered closely during the distillation with a piece of blotting paper or its equivalent, cut so as to fit the condenser tube tightly.

Procedure

4. (*a*) When everything is in readiness, heat shall be applied at a uniform rate, so regulated that the first drop of condensate falls from the condenser in not less than 5 nor more than 10 minutes. The distillation thermometer shall be read 2 minutes after heat is applied and the indication recorded as the *correction temperature*. This figure is of significance only in cases when there is a question as to the accuracy[6] of the initial boiling point, as subsequently determined. When the first drop falls

[6] The *initial boiling point* is one of the observations which it is customary to report as indicative of the quality of gasoline. It is a point which is very difficult to check, especially when duplicate tests are made at different room temperatures, and it has not been found practicable to devise an accurate method of correcting for this unavoidable variation in operating conditions. The practical significance of initial boiling point is not regarded as sufficient to warrant requiring that tests be conducted under regulated conditions of room temperature. The following directions are included as an expedient intended to obviate some of the difficulties that develop on account of the practice of including initial boiling point limits in specifications for the purchase and sale of gasoline:

If the *correction temperature* is below 70°F (21°C) or above 80°F (27°C), the observed initial boiling point, if it is below 150°F (66°C) may be considered incorrect by an amount not less than one third the difference between the *correction temperature* and 75°F (24°C). In case there is a dispute between the purchaser and the seller, the observed initial boiling point figures obtained by all parties shall be revised according to the following formula:

$$R = O - \frac{\text{Corrected Temperature (degrees Fahrenheit)} - 75}{3}$$

where R = the revised initial boiling point, and
O = the observed initial boiling point.

If all of the revised figures fall within the specification limit, the gasoline in question shall be considered as passing; otherwise, arrangements shall, if possible, be made to conduct a distillation test with the room temperature maintained between the limits of 70° and 80°F (21° and 27°C), inclusive.

from the end of the condenser, the reading of the distillation thermometer shall be recorded as the *initial boiling point.*

(*b*) The receiving cylinder shall then be moved so that the end of the condenser tube shall touch the side of the cylinder. The heat shall then be so regulated that the distillation proceeds at a uniform rate of not less than 4 ml per minute nor more than 5 ml per minute. The volume of distillate collected in the cylinder shall be observed and recorded, to the nearest 0.5 ml, when the mercury of the thermometer reaches each point that is a multiple of 10°C or the Fahrenheit equivalent of this point (30°C, 40°C, 50°C, 60°C, etc., or 86°F, 104°F, 122°F, 140°F, etc.). If preferred, the reading of the distillation thermometer may be observed and recorded when the level of the distillate reaches each 10-ml mark on the graduate. In case a product is being tested to ascertain whether or not it conforms with a given specification, all necessary observations shall be made and recorded, whether or not they are included in the series ordinarily employed by the laboratory making the test.

(*c*) No adjustment of the heat shall be made after the liquid residue in the flask is approximately 5 ml, unless the time required to bring over the last 5 ml of distillate and reach the end point exceeds 5 minutes. The *end point* is the maximum temperature observed on the distillation thermometer and is usually reached after the bottom of the flask has become dry.[7] If the bottom of the flask is not dry the operator shall record this fact.

(*d*) In case the time required to bring over the last 5 ml of distillate and reach the end point exceeds 5 minutes, the test shall be repeated and the heat shall be adjusted when the liquid residue reaches 5 ml. This adjustment may be either an increase or a decrease but must accomplish the purpose of bringing the period required to vaporize the last 5 ml of distillate and reach the end point within the limits of 3 and 5 minutes.

(*e*) The total volume of the distillate collected in the receiving graduate shall be observed at the conclusion of the distilla-

[7] When distilling certain naphthas manufactured for specific purposes, such as for use in paints and varnishes, the temperature indicated by the distillation thermometer when the last drop evaporates from the lowest point in the flask shall be recorded as the *dry point*. Any liquid hanging to the side of the flask or the thermometer should be disregarded.

tion and at not less than 2-minute intervals, until successive observations agree. This volume shall then be recorded as the *recovery*.

(*f*) The cooled residue shall be poured from the flask into a small cylinder graduated in 0.1 ml, measured when cool, and the volume recorded as *residue*.

(*g*) The sum of the recovery and the residue shall be calculated and recorded as *total recovery*.

(*h*) The difference between 100 ml and the total recovery shall be recorded as *distillation loss*.[8]

(*i*) The sum of the volume collected in the cylinder at any specified temperature, including the initial boiling point, and

TABLE 1. Values of the Constants "A" and "B" Used in Obtaining Corrected Distillation Loss

Observed Barometric Pressure, mm	*A*	*B*	Observed Barometric Pressure, mm	*A*	*B*
560	0.231	0.384	660	0.375	0.312
570	0.240	0.380	670	0.400	0.300
580	0.250	0.375	680	0.428	0.286
590	0.261	0.369	690	0.461	0.269
600	0.273	0.363	700	0.500	0.250
610	0.286	0.357	710	0.545	0.227
620	0.300	0.350	720	0.600	0.200
630	0.316	0.342	730	0.667	0.166
640	0.333	0.333	740	0.750	0.125
650	0.353	0.323	750	0.857	0.071
			760	1.000	0.000

[8] Abnormal distillation losses, when caused by low barometric pressure, may be corrected by the equation given below to an approximation of the loss to be expected under normal pressure. The corrected losses shall not be used in the calculation of percentages evaporated.

$$\text{Corrected loss} = AL + B$$

Where:
L = observed distillation loss, and
A and B = empirical constants, as given in Table 1 for various values of the observed barometric pressure.

the distillation loss may be recorded as the percentage evaporated at the temperature in question. Clear distinction shall be made between the inclusion or exclusion of the distillation-loss figure in the report of the volume of distillate obtained at any prescribed temperature. In the first case, the figure shall be reported as the volume *evaporated,* in the second, as the volume *recovered.*

Correction of Temperature Points to Standard Atmospheric Pressure

5. The actual barometric pressure shall be ascertained and recorded, but no correction shall be made except in case of dispute. In such cases, the temperature points shall be corrected to 760 mm (29.92 in.), by the use of the Sydney Young equation, as follows:

For Centigrade readings:

$$C_c = 0.00012\ (760 - P)\ (273 + t_c)$$

For Fahrenheit readings:

$$C_f = 0.00012\ (760 - P)\ (460 + t_f)$$

where C_c and C_f = the corrections to be added to the observed temperature t_c or t_f respectively, and P = the actual barometric pressure in millimeters of mercury.

Table 2 is a convenient approximation of the corrections as calculated by the above equations for those cases where the temperature for a specified volume is required.

Precision

6. The following criteria, based on 95 per cent probability, should be used for judging the acceptability of results:

(*a*) Duplicate results obtained by the same operator and apparatus should be considered suspect if they differ by more than the repeatability which, according to Figure 3, is appropriate to the test result obtained and to the rate of change in thermometer reading which prevailed at the stage at which the result was obtained.

(*b*) The results obtained by each of two laboratories should be considered suspect if they differ by more than the reproducibility which, according to Figure 3, is appropriate to the test result obtained and to the rate of change in thermometer reading which prevailed at the stage at which the result was obtained.

TABLE 2

Temperature Range °C	Temperature Range °F	Correction * per 10-mm Difference in Pressure °C	Correction * per 10-mm Difference in Pressure °F
10 to 30	50 to 86	0.35	0.63
30 to 50	86 to 122	0.38	0.68
50 to 70	122 to 158	0.40	0.72
70 to 90	158 to 194	0.42	0.76
90 to 110	194 to 230	0.45	0.81
110 to 130	230 to 266	0.47	0.85
130 to 150	266 to 302	0.50	0.89
150 to 170	302 to 338	0.52	0.94
170 to 190	338 to 374	0.54	0.98
190 to 210	374 to 410	0.57	1.02
210 to 230	410 to 446	0.59	1.06
230 to 250	446 to 482	0.62	1.11
250 to 270	482 to 518	0.64	1.15
270 to 290	518 to 554	0.66	1.19
290 to 310	554 to 590	0.69	1.24
310 to 330	590 to 626	0.71	1.28
330 to 350	626 to 662	0.74	1.32
350 to 370	662 to 698	0.76	1.37
370 to 390	698 to 734	0.78	1.41
390 to 410	734 to 770	0.81	1.45

* To be added if barometric pressure is below 760 mm; to be subtracted if barometric pressure is above 760 mm.

(*c*) To facilitate the use of Figure 3, the rate of change in thermometer reading in degrees Fahrenheit per the percentage evaporated or recovered, at any point between the initial boiling point and the end point or dry point, should be assumed to be the same as the average rate between two data points which are equidistant above and below the point in question, provided that the span from the point in question to either of the other data points does not represent more than 10 per cent evaporated or

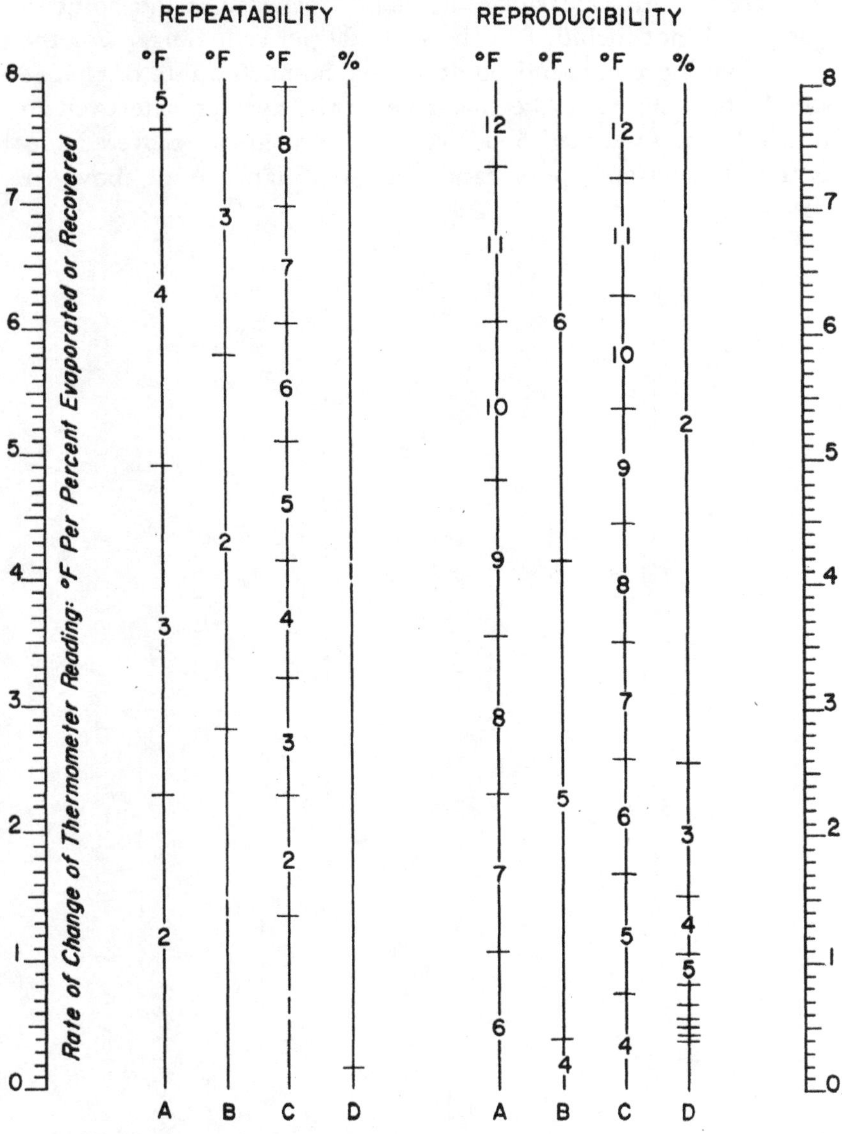

A. Initial Boiling Point
B. End Point or Dry Point
C. Thermometer Reading at Prescribed Percent Evaporated or Recovered
D. Percent Evaporated or Recovered at Prescribed Thermometer Reading

FIGURE 3. Precision of Distillation Method D 86

recovered in any case, nor more than 5 per cent if the point in question is not included in the 10 to 90 per cent range. For the initial boiling point, end point, or dry point, the rate of change should be assumed to be the same as the average rate over an interval, not to exceed 5 per cent evaporated or recovered, between the extreme point and the next data point above or below it.

VACUUM DISTILLATION

Although this method is quite similar to the one described in ASTM D 1160, the two will not always give the same result.

In this version changes have been made in the apparatus. It is now permissible to have the usual assembly of flask, condenser, and receiver fitted together by means of ball joints. It is emphasized that this is an alternate and existing items need not be discarded.

The method is applicable to petroleum products having an initial boiling point not lower than 200°F at 10 mm mercury absolute pressure.

Apparatus

(1) *Flask* similar in design to the 250-ml Saybolt distillation flask, but having a vapor outlet with an internal diameter of 8 ± 0.5 mm. Three different types are available:

- (a) made of Pyrex glass, with a wall thickness somewhat greater than normal to prevent distortion while in service
- (b) made of Vycor glass, as shown in Figure 1
- (c) made of Pyrex glass and fitted with ball joints at the ends of the neck and the vapor tube. This item includes an adapter for fitting the thermometer into the neck.

(2) *Shield* consisting of two pieces, as shown in Figure 1. The shield has a window, to permit the observation of the lower part of the neck during the test.

(3) *Condenser* consisting of an outer jacket made of Pyrex glass and a cooling tube of copper. This design permits rapid drainage of viscous products. Two different types are available. The essential difference is that one is fitted with ball joints and the other is not. The condenser illustrated in Figure 1 does not have the joints.

(4) *Receiver* consisting of a 100-ml glass-stoppered graduate mounted in a 5 × 9 in. battery jar. Satisfactory graduates are available with or without the ball joint.

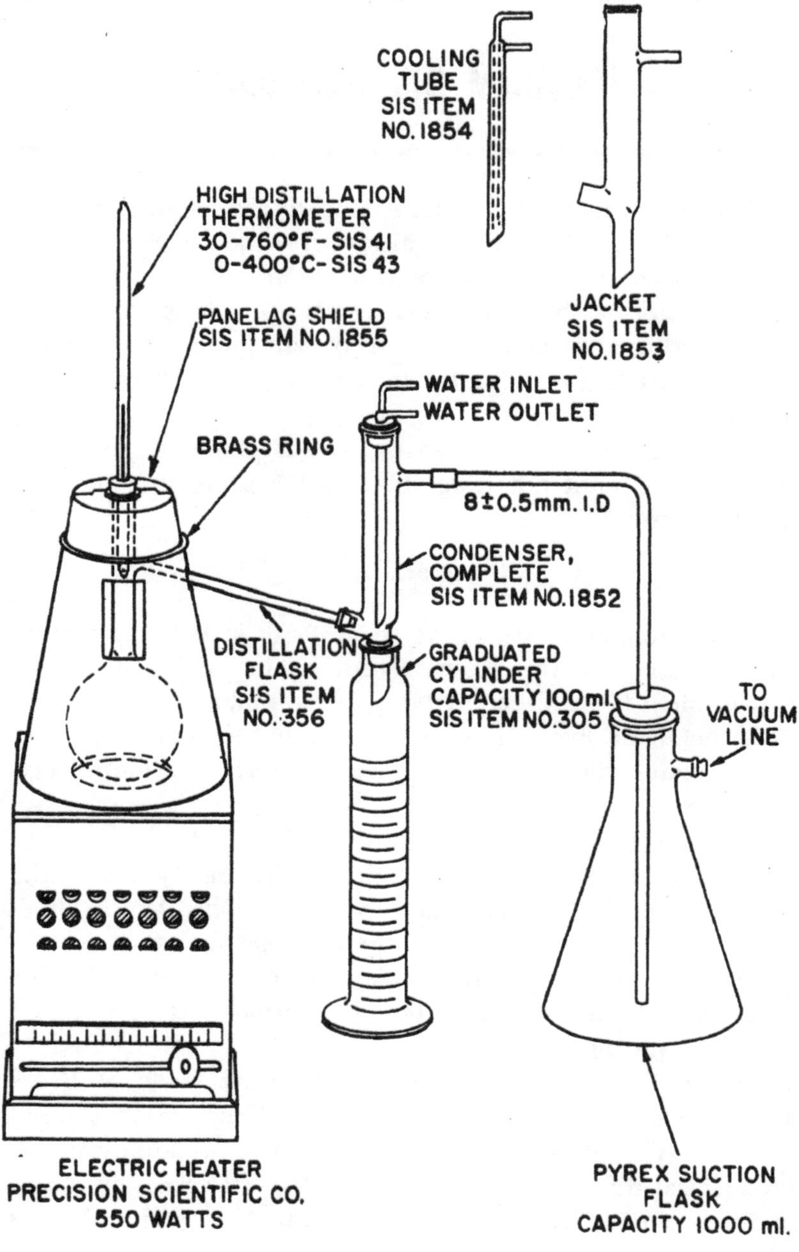

FIGURE 1. Modified ASTM Vacuum-Distillation Apparatus

In some cases, it is necessary to keep the water in the battery jar at a predetermined constant temperature. This may be done in any convenient manner. If desired, the jar may be sealed to the graduate above the 100-ml mark so that the jar and graduate are one piece of apparatus. In this case, the diameter of the jar may be reduced.

(5) *Electric Heater* of 550 watts capacity, with variable heat control. When the apparatus is assembled, a Transite board, measuring $5 \times 5 \times \frac{1}{4}$ in., with a central hole of $2\frac{3}{4}$ in. diameter, should be placed between the flask and the heater.

(6) *Vacuum Assembly*—Means should be provided for maintaining a 10 ± 0.2 mm absolute pressure inside the flask.

(7) *Thermometers*—Two different kinds will be needed: the regular ASTM high-distillation thermometer, and a general-purpose thermometer with a range of 30° to 220°F.

(8) *Steel Wool* of No. 1 grade.

Procedure

(1) Put a wad of steel wool, weighing 2 to 3 g, into the flask, pulling the wool apart into strands and inserting the strands so that the bulb of the flask is filled with the loose wool.

As received, the steel wool is coated with an oil compound to prevent rusting. Just before use, the wool should be rinsed thoroughly with light naphtha to remove this coating and then dried. The wool inside a flask may be used more than once providing that it is rinsed and dried between tests. It is added to prevent frothing and bumping of the sample.

(2) Measure 100 ml of the sample as received into the graduate and transfer the oil to the distillation flask. Allow the graduate to drain as completely as possible and then, without rinsing, attach it to the condenser.

In some cases, it may be desirable to heat the sample to insure homogeneity and to pour it into the graduate. If heating is necessary, fill the graduate with, and note the temperature of, the hot oil. In either case, allow the graduate to drain until not more than 1 ml remains in the cylinder, collecting all of the drainings in the flask.

Occasionally the *front-end volatility* (i.e., the amount of the lower-boiling fractions) is of critical importance. In this case, it is recommended that a clean graduate should be used to collect the distillate.

(3) Put another wad of steel wool, weighing 0.15 to 0.30 g, into the neck of the flask and push it down to a point just above the junction of the neck and the bulb of the flask. Put the thermometer into place, locating it so that the bulb is in the middle of the neck and the bottom of the capillary is on a level with the inside of the bottom of the vapor tube.

The extra wad of steel wool helps prevent the liquid from frothing up into the neck of the flask. When the flask with the plain neck is used, the thermometer should be mounted in a cork and the cork sealed to the flask with sodium silicate solution. When the flask with the ball joint is used, the thermometer may be held in place either by cementing it into the adapter with some satisfactory cement, such as litharge and glycerin, or by slipping a section of rubber tubing over the neck of the adapter and sliding the thermometer through the tubing.

(4) Put the charged flask on top of the electric heater, with a Transite board located between the two. Put the shield around the flask, as shown in Figure 1. Although not mandatory, it is suggested that some sort of reinforced glass screen is placed around the assembly. This will minimize danger from fragments if the flask should break.

(5) Fill the battery jar with water, insert the graduate with the attached condenser, and connect the bottom of the condenser jacket to the discharge arm of the flask. Attach the upper arm of the jacket to a glass tube leading to the bottom of the suction flask having a capacity of at least 1 liter. The suction flask should be loosely filled with cotton or glass wool.

The temperature of the water in the battery jar must be the same as that of the sample at the time the 100-ml portion was measured. If it was necessary to heat the sample originally, the water must be kept at the same temperature.

(6) Attach the side arm of the suction flask to the source of vacuum by means of heavy-walled rubber tubing. The line should contain a T- or Y-joint, with the side arm of the T attached to a vacuum gage. Attach the inner tube of the condenser to the water line. When the apparatus is finally assembled, seal all of the joints with some satisfactory cement, such as water glass

or a mixture of 75 per cent collodion and 25 per cent castor oil. (This sealing may be omitted on all of the ball joints.)

(7) When everything is ready, apply suction to the flask until a pressure of 10 ± 0.2 mm is obtained, and keep the system under this pressure throughout the test. If necessary, turn on the condensing water to the condenser.

(8) When the absolute pressure reaches 10 mm, apply heat to the flask. This must be done gradually to prevent foaming of the sample. When the first drop falls into the condenser, note the temperature on the thermometer and record it as the *initial boiling point.* The time between the application of the heat and the falling of the first drop should be not less than 20 nor more than 30 minutes.

(9) After the first drop has fallen, regulate the heat so that the distillation will proceed at a uniform rate of 4 to 5 ml per minute throughout the test. Continue the distillation, recording the temperature on the thermometer whenever the volume of liquid in the graduate is an even multiple of 10 per cent, or at any specified point.

The operator must find, by trial and error, the proper heating rate to keep the distillation going evenly. By the time the volume of condensate reaches 15 ml, the sample should be distilling at the rate of 4 to 5 ml per minute.

(10) Continue the distillation until the temperature, as noted on the thermometer, reaches a maximum and starts to fall, or until the temperature reaches 620°F, whichever occurs first. (Do not continue heating beyond 620°F.) When the heating is stopped, note and record the maximum temperature reached on the thermometer. Allow the system to drain for a few moments, and record the volume of liquid in the graduate. After cooling, slowly bring the system to atmospheric pressure.

Report the initial boiling point, the percentage recovered at each even multiple of 10 per cent (or other specified points), the final boiling point, and the total amount recovered. If the test was stopped at 620°F, substitute "Stopped at 620" for the final boiling point.

With due care and attention to details, duplicate results, at all points except the initial, should not differ from the mean by more than 5 per cent of the mean.

The essential feature of this version of vacuum distillation is the use of glassware which is fitted with ball joints. In routine laboratories, where this test is frequently carried out, the total output can be increased materially without affecting the precision of the results.

In earlier versions, the shield was made of Sil-O-Cel and was in three pieces. The Sil-O-Cel shields were rather soft and quite easily broken. Paneleg shields, as shown in Figure 1, are much less fragile. However, any shields of the Sil-O-Cel type which are still in use need not be discarded.

DISTILLATION OF AROMATIC HYDROCARBONS

This method is the same as that published by the ASTM under their designation D 850–55.

It is applicable to industrial aromatic hydrocarbons of relatively narrow boiling ranges. The initial boiling point should be higher than 30°C, and the dry point below 250°C.

Apparatus

(1) *Flask* of 200-ml capacity, as shown in Figure 1.

(2) *Thermometers* graduated in intervals of 0.2°C (except 42°C, which is 0.5) with the following ranges:

Range, °C	
48 to 102	ASTM No. 39 C.
72 to 126	ASTM No. 40 C.
98 to 152	ASTM No. 41 C.
95 to 255	ASTM No. 42 C.
70 to 120	Barrett
110 to 160	Barrett

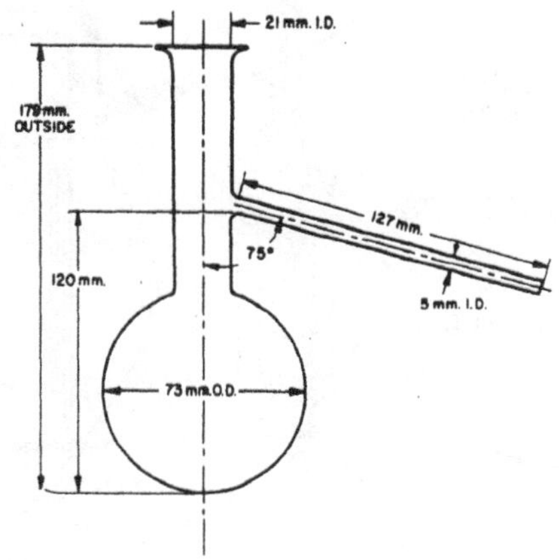

FIGURE 1. Dimensions of Distillation Flask

Note that the ASTM thermometers are all of the partial-immersion type, and corrections for emergent stem are not necessary. The Barrett thermometers are of the total-immersion type and such corrections *must be applied* to the observed readings. Comparable results can be obtained with either type. To simplify the procedure, it is recommended that the ASTM thermometers should be used for all products tested by this method and the Barrett thermometers should be used only when definitely specified by the customer. In this case, another thermometer will be required for determining the temperature of the exposed portion of the stem of the Barrett thermometer, and the ASTM general-purpose thermometer of the range of 0° to 105°C will be satisfactory for this purpose. In any case, all thermometers should be checked for accuracy.

(3) *Condenser Assembly*—Either of two assemblies may be used. The preferred type is the familiar Engler apparatus as described and illustrated under L.I.C. No. 10.01H, which should be used whenever practicable. The arrangement shown in Figure 2 *must* be used whenever the sample under test contains naphthalene or other products which are solid at a temperature

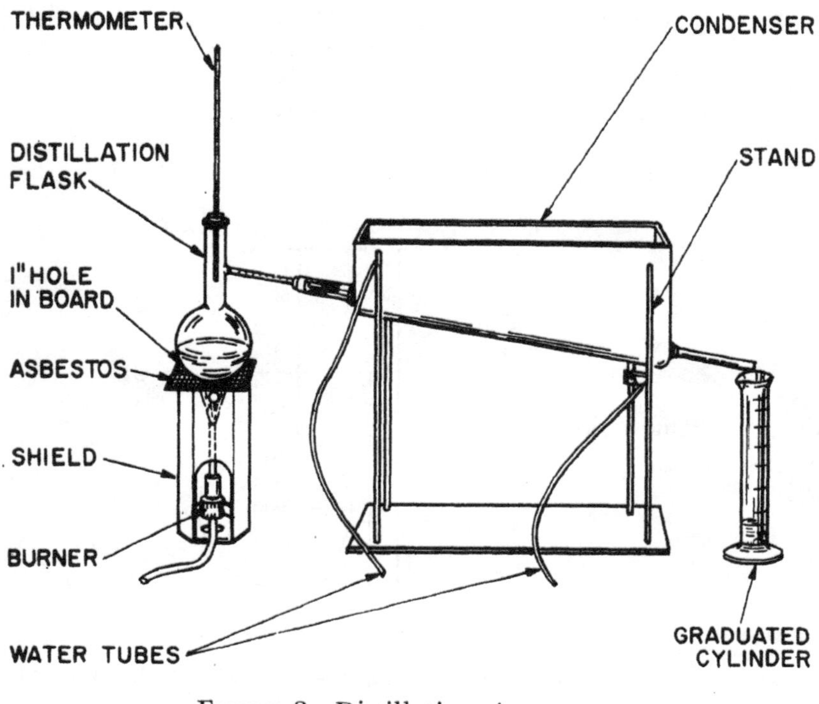

FIGURE 2. Distillation Apparatus

higher than +5°C. If such products are present, the operator can notice the early formation of solid crystals in the glass tube and stop the heating before the tube is completely plugged.

In the assembly of Figure 2, the condenser is a straight glass tube, 605 ± 5 mm in length and of 12 mm inside diameter. The tube is of standard wall thickness (about 1.25 mm) with the exit end cut off square and ground flat. It is mounted in a cooling trough at an angle of 75° from the vertical and in such a manner that at least 15 in. of the tube is in contact with the cooling water and the clearance between the tube and any parallel side of the trough is at least 0.75 in. The water in the trough should be maintained at a temperature of 15 ± 5°C, either by adding ice or by circulating cold water.

The assembly of tube, trough, and stand may be obtained from the Emil Greiner Co., 20 N. Moore Street, New York City, their designation being G–1975. A permissible assembly can be made locally by removing the metal tube from the condenser prescribed in L.I.C. No. 10.01H and replacing it with a glass tube conforming to the given dimensions.

The flask may be heated by either an electrical heater or a gas flame. If the heater is used, it should be fully adjustable and capable of giving sufficient heat to distill the product at the desired range. If a gas burner is used, it must be mounted inside the shield shown in Figure 2 and adjusted to produce a completely blue flame. The top of the shield should be about 2 in. higher than that of the burner.

(4) *Accessories*—The following items will also be needed: a graduated receiver of 100-ml capacity and asbestos boards of 6 × 6 × ¼ in., with 1-in. central hole.
with 1½-in. hole.
with 2-in. hole.

Preparation of the Sample

Before starting the test, be sure that the sample is transparent and free of separated water. Any separated water can normally be removed by care in pouring the 100-ml portion into the cylinder. If necessary, water or cloudiness can be removed by filtration. This should be done in a draft-free area, and the operator should use a soft paper through which the sample will filter rapidly. The funnel should be covered with a watch glass, and at least 200 ml of the sample filtered. Dehydration (removal of dissolved water) is not permissible.

Procedure

(1) Select the proper thermometer and asbestos board for the sample under test. The choice depends on the boiling range of the sample, as shown in Table 1.

TABLE 1

Product	Thermometer	Board
Benzene	ASTM 40 C	1-in. hole
Toluene	ASTM 40 C	1½-in. hole
Xylene	ASTM 41 C	1½-in. hole
Dry point above 145°C	ASTM 42 C	2-in. hole

(2) Measure 100 ml of the sample at room temperature into the graduate and pour the liquid into the flask. Allow the graduate to drain for 15 seconds and then mount it under the end of the condenser. Put the proper thermometer, mounted in a cork stopper as shown in Figure 2, into the mouth of the flask, locating it so that the top of the bulb (or the top of the contraction chamber, if present) is level with the lowest point of the junction between the side tube and the neck of the flask. Make sure that the thermometer is vertical and centrally located in the neck of the flask. Put the proper asbestos board into place and mount the flask on it in such manner that the end of the side arm will extend through a cork and into the condenser for a distance of about 2 in.

(3) If corrections for emergent stem are required, mount another thermometer beside the one in the flask, locating the second one so that its bulb will be about halfway between the top of the cork in the flask and the expected dry point.

(4) When everything is ready, apply heat to the flask and start a stopwatch or timer. Heat the flask slowly, especially after boiling has started, so as to allow the mercury column of the thermometer to become fully expanded before the first drop distills over. Regulate the rate of heating so that the ring of

condensing vapor on the wall of the flask reaches the lower edge of the side arm in not less than 90, and preferably 120, seconds from the start of the rise of the vapor ring.

(5) When the first drop falls into the receiver, note and record the temperature and the time since the start of the test. The total time from the start of the heating until the first drop falls should be not less than 5 nor more than 10 minutes.

(6) When distillation starts, move the receiver to allow the condensate to flow down its inner wall without spattering. Adjust the heater to continue the distillation at a rate of 5 to 5.5 ml per minute.

The ASTM method allows a rate of 5 to 7 ml per minute. Experience has shown that better precision can be obtained by controlling the rate more closely. At all times, the distillation should proceed at an even, steady rate without major changes in the application of heat.

(7) During the distillation, record the reading of the thermometer when 5 per cent, 10 per cent, and each additional 10 per cent up to 90 per cent, and 95 per cent of the sample have distilled over. Take the final reading when the last of the liquid just disappears from the bottom of the flask. Record the last temperature as the *dry point.*

When testing crudes the bottom of the flask may not become completely dry before the temperature starts to fall. In this case, record the maximum temperature, noted on the thermometer, as the *decomposition point* of the sample. The decomposition point may also be indicated by the appearance of heavy fumes in the flask. Should this occur, note and record the temperature when the bulb of the flask becomes essentially full of fumes.

(8) After the dry (or decomposition) point has been recorded, stop the heating and note the total yield of distillate in the receiver. The recovery should be at least 97 per cent for close-boiling solvents, such as toluene and xylene, and at least 95 per cent for other liquids. If this recovery is not obtained, repeat the test.

(9) Also record the following data:

(a) the number of degree graduations on the emergent stem from the top of the cork to the graduation at the observed temperature

(b) the reading of the second thermometer (if used), representing the average temperature of the exposed stem

[Items (a) and (b) are needed only when the Barrett thermometer is used.]

(c) the barometer reading and the temperature of the barometer; the observed barometric pressure should be corrected to millimeters of mercury at 0°C

(d) the correction for inaccuracy of the thermometer.

Temperature Corrections

If required, calculate the emergent stem correction by the equation:

$$C = K \times N \times (T - t)$$

where

K is the differential-expansion coefficient of the glass and the mercury in the thermometer. This may be taken as 0.000154.

N is the number of degree graduations exposed on the stem.

T is the reading of the distillation thermometer in Centigrade degrees.

t is the reading of the auxiliary thermometer in Centigrade degrees.

C is the correction, in Centigrade degrees, which is to be added to the first reading.

After this correction has been made, make a further correction for variation from standard barometric pressure by the equation:

$$C = K(760 - P)$$

where

P is the observed barometric pressure, corrected to 0°C.

C is the correction to be applied.

K is the constant which may be found in Table 2.

The correction is to be added to the observed temperature when the corrected barometric pressure is *below* 760 mm and subtracted when the corrected pressure is above 760 mm.

TABLE 2

Liquid	K
Benzene	0.043
Toluene	0.046
Ethylbenzene	0.049
Orthoxylene	0.050
Metaxlyene	0.049
Paraxylene	0.049
Nitration-Grade Xylene, 5 or 10°	0.049
Solvent Naphtha	0.049
Hi-flash Naphtha	0.053
Mixed Xylenes	0.048
Aromatic Solvent Naphthas	0.049

The ASTM methods 850 and 1078 are essentially identical; the only differences are in the choice of thermometers and the optional use of the glass condenser. As previously stated, this condenser should be used only when the product contains some compound which might solidify in the tube. As for the thermometers, there may be some doubt as to whether the operator did or did not apply the necessary stem correction when the Barrett thermometer was used. If ASTM thermometers are used, there will be no question whatever on this subject.

DETERMINATION OF HIGH-BOILING FRACTIONS IN ISOPROPYL ETHER

This method may be used on either crude or redistilled isopropyl ether.

Apparatus

(1) *Fractionating Column*—A Fenske column, 3-ft long with an internal diameter of 1 in., fitted with a chilled-water distillation head, should be used. The column may be either heated electrically or fitted with a vacuum jacket, the second being preferred. Normally, the column is made in the laboratory. If desired, however, a satisfactory column may be obtained from the Todd Scientific Co., Springfield, Pa.

If the apparatus is to be assembled by laboratory personnel, the following suggestions will be helpful. A satisfactory column can be obtained from the Ace Glass Co., Vineland, N.J., under their catalog number 6585. The column should be packed with helical packing of $\frac{3}{32}$ in. internal diameter, made from No. 30 B & S gage Nirosta wire, obtainable from Dr. M. R. Fenske, Petroleum Refining Laboratory, Pennsylvania State College, State College, Pa. A satisfactory distillation head of modified Booty type is available at the Ace Glass Co. This head should contain a thermometer (Ace Glass Co. catalog number 8315) with a 10/30 ST joint, range 0° to 110°C, with 3 in. of stem below the joint.

Procedure

(1) Measure 250 ml of the sample and 150 ml of water into a 1-liter round-bottom flask. The flask should be attached to the column and the sample refluxed for 10 minutes before removing any distillate.

(2) The distillation should then be carried out at a 5 to 1 reflux ratio until the temperature of the distillate reaches 90°C, collecting the condensate in any convenient receiver. When the temperature reaches 90°C, the receiver should be removed and replaced with an accurate 25-ml graduate.

(3) Distillation should continue until the volume of organic condensate in the graduate remains constant. Then distillation should be stopped and the volume of the higher-boiling fractions in the graduate recorded. This volume should be calculated to percentage and so reported.

TEST FOR LIGHT ENDS IN ISOBUTYLENE AND SIMILAR PRODUCTS

This method is based on a similar one appearing in the *Laboratory Manual* of the Rubber Reserve Corp. under the designation LM 7.11.4. It is suitable for determining the amount of matter boiling below −15°C in C_4 fractions and related mixtures.

A large portion of the sample under test is charged into a vacuum-jacketed fractionating column similar to the one shown in Figure 1. The distillation is conducted at low temperatures, and the gases which distill over are collected in one or more special receivers. Generally, about 70 per cent of the total recovery is taken from the vapor line, and the remainder from the liquid phase. The recovered gases are then charged to a Podbielniak column and analyzed by conventional procedure. The percentage of light ends in the recovered gas is calculated back to the original-sample basis.

Apparatus

(1) *Distillation Assembly* as shown in Figure 1, consisting essentially of a flask, a vacuum-jacketed column, a special still head, and a receiving unit, all connected by ground glass joints. The still head consists of a reflux condenser, a side arm for a thermometer or thermocouple, and a central chamber containing a movable funnel which is adjacent to a magnetic coil. Depending on the location of the funnel, the condensate from the upper condenser will drop either into the main column or into the line leading to the receiver. Thus, by attaching a timer to the coil, the reflux ratio of the distillate can be controlled at will. Satisfactory parts can be obtained from either the Ace Glass Company, Vineland, N.J., or the Glass Engineering Laboratories, Belmont, Cal., under the following numbers:

Description	Ace Catalog No.	Glass Eng. Lab. Catalog No.
Automatic still head, including reflux condenser, movable funnel, and overhead connections, but without thermometer, coil unit, or timer	6530	Serial D-5
Reflux ratio timer	6560	D-40
Coil unit	6555	D-25
Vacuum-jacketed column	6585	Not available

The distilling flask is not listed, but may be obtained from either firm on special order. The volume must be specified, and depends on the expected amount of light ends in the sample. In most cases, a sample charge of 2500 g is used; the volume of the flask should, therefore, be 4 liters. In the flask shown, there is a thermometer well. The well is to be used for the introduction of the sample. The order should, therefore, specify that the lower end of the well is to be left open, and the upper (protruding) end fitted with a glass stopcock.

The column should be vacuum jacketed, with the walls silvered. While the exact dimensions are not mandatory, the overall length should be about 5 ft, and the internal diameter, $\frac{1}{2}$ in. The column is filled with $\frac{3}{32}$ in. helices made of glass or stainless steel.

The receiving unit shown in Figure 1 is suggested, not mandatory. Any convenient type may be used, providing that the vapors can be drawn into the gas-collection bottles described in the next section and that the receiver may be warmed if necessary to vaporize the last traces of liquid.

The temperature of the overhead fractions may be measured by a thermometer, as illustrated, or by a thermocouple. The reading range should cover $-15° \pm 10°C$.

Whatever the assembly, it is suggested that all of the parts are ordered from either one of the given firms. Items supplied by the Glass Engineering Laboratories will be fitted with standard taper joints, while those from Ace will have the spherical joints, as shown. Both are satisfactory.

(2) *Gas-Collection Assembly*—Provision must be made for collecting about 6 gal of gas from the column. Since a container of this size is awkward to handle, the gas is usually collected in

two 3-gal containers, which may be of glass, or preferably of corrosion-resistant metal.

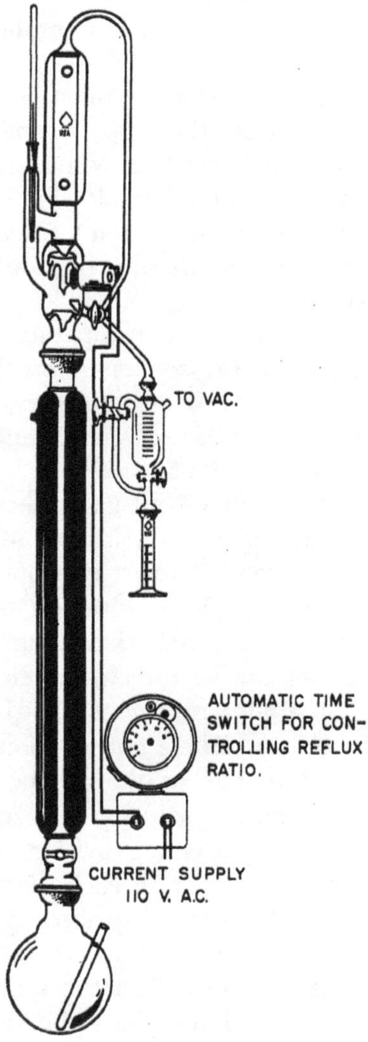

FIGURE 1. Distillation Apparatus for Low-Boiling Hydrocarbons

Metal containers of this size are not generally available commercially, but can be made easily in the local shops. These containers are cylindrical, tapered to a pointed cone at both ends, with the tip of each cone fitted with a metal valve. While the exact volume is immaterial, it must be known accurately.

(3) *Podbielniak Column*—Any of the available types may be

used. Although operating instructions usually accompany the instrument, complete details will be supplied on request.

Important Note

(1) The apparatus must be assembled in a place safe from any source of ignition. It is strongly recommended that the assembly is done in a room in which the use of gas burners or other open flames is forbidden. If this is not feasible, the column should be mounted in a hood fitted with doors and a strong draft. Under no circumstances should smoking be permitted near the assembly.

(2) Assemble the apparatus as shown in Figure 1. Wrap the outside of the condenser on the still head with some insulating material. Make all necessary provisions for circulating Varsol or other similar solvent through the condenser at a temperature of about $-15°C$.

(3) Fill two 3-gal gas-collection bottles with a saturated solution of magnesium sulfate. Mount these on the bench top near the receiver. Connect the tops to a T-joint, and connect the arm of the T to the receiver on the column. Make some provision, either with pinch clamps or stopcocks, that the gas from the receiver can be directed to either bottle.

(4) Mount two more 3-gal bottles (these may be of glass) near the others. These bottles should be empty. Each should be connected to one of the others in such a manner that the magnesium sulfate solution will siphon from one to the other as gas enters the system. Provision must be made for determining the volume of vapor in each of the original bottles after its liquid content has been displaced by the incoming gas.

Specific instructions at this point depend on the type of bottle used to collect the sample. If a metal container is used, the liquid should be siphoned into a graduated container; the volume of liquid collected is the same as that of the gas sample. If a glass container is used, the bottle may be marked on the side in increments of 500 ml, and the volume of gas read directly. In either case, it is important that the collected gas should not be under pressure, either positive or negative. In other words, the siphon should be so arranged that when the collection bottle is full of gas, the level of the displaced liquid in the second bottle will be about the same as the bottom of the collection bottle.

(5) When the apparatus is completely assembled, surround the flask at the bottom of the column with a dry ice–alcohol bath, and chill to −70°F. Circulate Varsol or mineral spirits at −15°C through the condenser.

Procedure

(1) When everything is ready, weigh the bomb containing the sample, as received, to the nearest gram. Attach to the flask, open the adjacent stopcocks, and allow a suitable amount of sample to flow into the flask. Close the stopcocks, weigh the bomb again, and record the difference as the weight of sample taken.

The weight of sample to be taken depends on the expected content of light ends. For samples containing up to 0.03 per cent by weight, a charge of 2500 g is recommended. For higher amounts of light fractions, the sample weight may be reduced proportionately.

CAUTION. Before the sample is introduced, turn the stopcock near the solenoid so that the air in the column will flow into the collection bottles. From here on, this stopcock must *always* be in such a position that the overhead fractions, either liquid or vapor, can pass into the receiver. If it is ever closed, a build-up of pressure could cause a violent explosion.

(2) Keep the apparatus under total reflux for about ½ hour, collecting any vapors in one of the gas-collection bottles. During this time, the apparatus should be arranged so that any collection will be only from the vapor phase of the overhead and not from the liquid.

(3) Gradually raise the temperature of the liquid in the flask, still collecting the overhead through the vapor line, until about 4 gal of gas have been collected in the bottles. Set the timer for a 3 to 1 reflux ratio and turn the stopcock so that the sample will be collected from the liquid phase. If necessary, warm the receiver so that any liquid in it will be vaporized and pass into the bottles. Continue the distillation until an additional 2 gal of

gas have been collected. Then stop the distillation and record the volume of gas in the bottles, the temperature of the room, and the barometric pressure.

(4) Charge the collected gas into a Podbielniak column and distill in the normal manner.

Calculations

(1) Calculate the results of the Podbielniak analysis to volume per cent, and then to weight per cent, assuming 42 and 56 as the molecular weights of the C_3 and C_4 fractions, respectively. Using these results, calculate the weight per cent of C_3 in the original sample.

Example—Suppose that the following results were obtained in the course of the test:

Total Original Charge	2500 g
Total Overhead in the First Distillation	22,710 ml (uncorrected)
Room Temperature	30°C
Barometric Pressure	765 mm

Podbielniak Analysis: 9.8 mol per cent of C_3, and 90.2 mol per cent of C_4

(a) The molecular weight of the mixed gas charged to the Podbielniak column would be

$$
\begin{aligned}
0.098 \times 42 &= 4.11 \\
0.902 \times 56 &= 50.50 \\
\hline
&\ 54.61 = \text{Molecular weight of the mixed gas}
\end{aligned}
$$

(b) The total weight, in grams, of the overhead gas would be:

$$\frac{22710 \times (765 - 20^{a}) \times 273 \times 54.61^{b}}{22412 \times 760 \times (273 + 30)} = 48.0 \text{ g}$$

where

a is the correction for water-vapor pressure
b is the molecular weight of the mixture

(c) The amount of C_3 in the overhead would be:

$\frac{(4.11 \times 100)}{54.61}$ or 7.5 per cent by weight of C_3 in the overhead

(d) The final equation then becomes:

$\frac{48.0 \times 0.075 \times 100}{2500}$ or 0.144 weight per cent of C_3 in the original sample

If a mass spectrometer is available, this analysis may be carried out by means of it. When the content of light ends is over 0.1 per cent, the sample may be charged directly into the spectrometer without a preliminary separation. When the content is less than this, the sample may be charged into a Podbielniak column and the proper fraction analyzed by the spectrometer. In either case, the preliminary distillation may be omitted if a mass spectrometer is available.

FLASH POINT BY THE TAG CLOSED TESTER (ABEL FLASH POINT)

The method of test recommended for determining flash points by the Tag (or Abel) closed tester is the same as that described in the current version of ASTM D 56, subject to the qualifications outlined in the next paragraph.

In earlier years, some specifications for solvents required an Abel flash point, rather than a Tag. The Abel instrument is no longer in common use. If an Abel flash point is required and the instrument is not available, the flash point may be determined in the Tag tester and, after subtracting 5°F, the corrected result may be reported as the Abel flash point. To show that the determination was actually made in the Tag tester, include the letter *t* in the report.

Example—If a sample showed a flash point of 133°F in the Tag tester, subtract 5° and report as *Abel flash point, 128t.*

If the determination is made in the Abel tester, include the letter *a,* to indicate this fact (Abel flash point, 128*a*).

The factor of 5° allows a margin of safety; in many cases, the actual Abel flash points are only 2 or 3° lower than the Tag flash points.

ASTM DESIGNATION: D 56–56

Scope

1. This method describes a test procedure for determining the flash point of all mobile liquids flashing below 175°F (79°C). The method is applicable to lacquer solvents and diluents of low flash point, but not to products classed as fuel oil.[1]

[1] For the determination of flash points of fuel oils, use ASTM Method D 93, Test for Flash Point by Means of the Pensky-Martens Closed Tester.

Outline of Method

2. The sample is placed in the cup of the tester and, with the lid closed, heated at a slow constant rate. A small flame of specified size is directed into the cup at regular intervals. The flash point is taken as the lowest temperature at which application of the test flame causes the vapor above the sample to ignite.

Apparatus

3. (*a*) *Tag Closed Tester*—The apparatus is illustrated in Figure 1 and described in detail in Appendix 1. Refer to Appendix 2 for directions for checking the condition and operation of the tester.

(*b*) *Shield*—A shield, 18 in. (46 cm) square and 24 in. (61 cm) high, open in front, is recommended.

(*c*) *Thermometers*[2]—For the test-cup thermometer, use one as prescribed in Table 1. For the bath thermometer, any convenient type which has an adequately open scale, covering the required range, may be used; it is often convenient to use the same type of thermometer as used in the test cup.

Sample

4. Erroneously high flash points may be obtained if precautions are not taken to avoid the loss of volatile matter. Containers shall not be opened unnecessarily and transfers shall not be made unless the sample temperature is at least 20°F (11°C) below the expected flash point. Samples in leaky containers shall be discarded.

Preparation of Apparatus

5. (*a*) Support the tester on a level steady table. Unless tests are made in a draft-free room or compartment, surround the

---

[2] Whenever thermometers complying with ASTM requirements are not available, thermometers complying with the requirements of The Institute of Petroleum, thermometer IP 15 F P.M.-Low may be used.

tester on three sides by the shield for protection from drafts. Tests made in a laboratory draft hood or near ventilators are not reliable.

(*b*) Gas is recommended for the test flame. If gas is not available, insert a wick of cotton in the burner tip, place a small quantity of cotton waste in the chamber to which the burner tip is attached, and fill the chamber with signal, sperm, or lard oil.

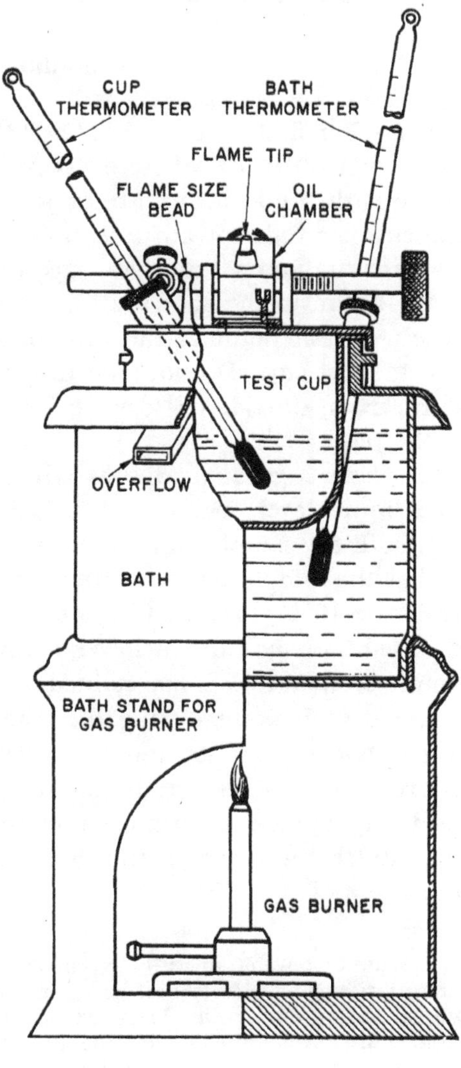

FIGURE 1. Tag Closed Flash Tester

TABLE 1. Thermometers

For Tests	Below 40°F (4°C)	At 40° to 120°F (4° to 49°C)	Above 120°F (49°C)
Use ASTM Thermometer	57°F or 57°C	9°F or 9°C 57°F or 57°C	9°F or 9°C

Procedure

6. (*a*) For flash points of 55°F (13°C) or higher, fill the bath with water until it starts to overflow. For lower flash points, use as the bath liquid a mixture of water and ethylene glycol (permanent-type radiator antifreeze) or any other suitable liquid of low viscosity and suitably low freezing point.[3] The temperature of the liquid in the bath shall be at least 20°F (11°C) below the expected flash point at the time of introduction of the sample into the test cup. Do not cool the bath liquid by direct contact with carbon dioxide or "dry ice." Place the test cup in position in the bath.

(*b*) Using a graduate and avoiding wetting the cup above the final liquid level, measure 50 ± 0.5 ml of the sample into the cup, both the sample and graduate being precooled, if necessary, so that the sample temperature at the time of measurement will be 60° ± 10°F (16° ± 5.6°C) or at least 20°F (11°C) below the expected flash point, whichever is lower. It is essential that the sample temperature is maintained at least 20°F (11°C) below the expected flash point during the transfers from the sample container to the graduate and from the graduate to the test cup. Destroy air bubbles on the surface of the sample. Wipe the inside of the cover with a clean cloth or absorbent tissue paper; then attach the lid, with the thermometer in place, to the bath collar.

[3] Because of possible difficulty in maintaining the prescribed rate of temperature rise and the formation of ice on the lid, results by this method for samples having flash points below 32°F (0°C) may be somewhat unreliable. Trouble from ice formation on the slide may be minimized by carefully lubricating the slide shutter with high-vacuum silicone lubricant.

(*c*) Light the test flame, adjusting it to the size of the small bead on the cover. Adjust the heat so that the temperature of the sample will rise at a rate of 1°F (0.6°C) each 30 ± 3 seconds.

(*d*) When the temperature of the sample in the test cup is 10° F (5.6°C) below its expected flash point, operate the mechanism on the cover in such a manner as to introduce the test flame into the vapor space of the cup and immediately bring it up again. The time consumed for the full operation shall be about 1 second, or the time required to pronounce distinctly the words "thousand and one." Avoid any jerkiness in the operation of depressing and raising the test flame.

(*e*) Repeat the application of the test flame after each 1°F (0.6°C) rise in temperature of the sample in the test cup, discontinuing the test and removing the source of heat when a distinct flash is observed in the interior of the cup. Do not confuse the true flash with the bluish halo which sometimes surrounds the test flame during applications immediately preceding the actual flash. Record the temperature of the sample when the flash point is reached.

(*f*) Lift the lid and wipe off the thermometer bulb. Remove the sample cup, empty, and wipe dry.

(*g*) If, at any time between the first introduction of the test flame and the observation of the flash point, the rise in temperature of the sample is not within the range 1°F (0.6°C) each 30 ± 3 seconds, or if the actual flash point differs from the expected flash point by an amount greater than the applicable repeatability limit in Section 8, discard the result and repeat the test,[4] adjusting the source of heat to secure the proper rate of temperature rise and/or using a modified "expected flash point," as required.

Correction for Barometric Pressure

7. Observe and record the barometric pressure at the time of the tests. When the pressure differs from 760 mm of mercury, correct the flash point by means of the following equation, the corrected flash point being recorded to the nearest whole number.

---

[4] Never make a repeat test on the same portion of sample once used; always take a fresh portion of sample for each test.

$$\text{Corrected flash point} = F + 0.06\,(760 - P)$$

where:

F = observed flash point, in Fahrenheit degrees, and
P = barometric pressure, in millimeters of mercury.

Precision

8. The following data should be used for judging the acceptability of results (95 per cent probability).

(*a*) Duplicate results by the same operator should be considered suspect if they differ by more than 2°F (1.11°C).

(*b*) The result submitted by one laboratory should not be considered suspect unless it differs from that of another laboratory by more than the following amounts:

Flash Point	Reproducibility
Below 55°F (12.8°C)	6°F (3.3°C)
55°F (12.8°C) or above	4°F (2.2°C)

APPENDIX 1

Apparatus

The Tag closed tester shall consist of the test cup, lid with test flame, and liquid bath conforming to the following requirements:

A1. *Test Cup,* of brass or other nonrusting metal of equivalent heat conductivity and conforming to the dimensional requirements prescribed in Table 2. It shall weigh 68 ± 1 g.

TABLE 2. Dimensional Requirements

	In inches
Depth of bath liquid surface below top of test cup	1.094 ± 0.016
Depth of sample surface below top of test cup	1.156 ± 0.031
Depth of bottom of bulb of test thermometer below top of cup when in place	1.77 ± 0.03
Inside diameter of test cup at top	2.125 ± 0.005
Diameter of bead on top of cover	0.156 ± 0.031
Diameter of opening in tip of test flame nozzle	0.049 ± 0.010
Outside diameter of tip of test flame nozzle	0.079 max.

A2. *Lid*—(*a*) The lid comprises a circle of nonrusting metal with a rim projecting downward about $\frac{5}{8}$ in. (1.59 cm), a slide shutter, a device which simultaneously opens the shutter and depresses the tip of the tube which carries fuel through to the test flame, and a slanting collar in which the cup-thermometer ferrule is inserted. Figure 2 gives a diagram of the upper surface

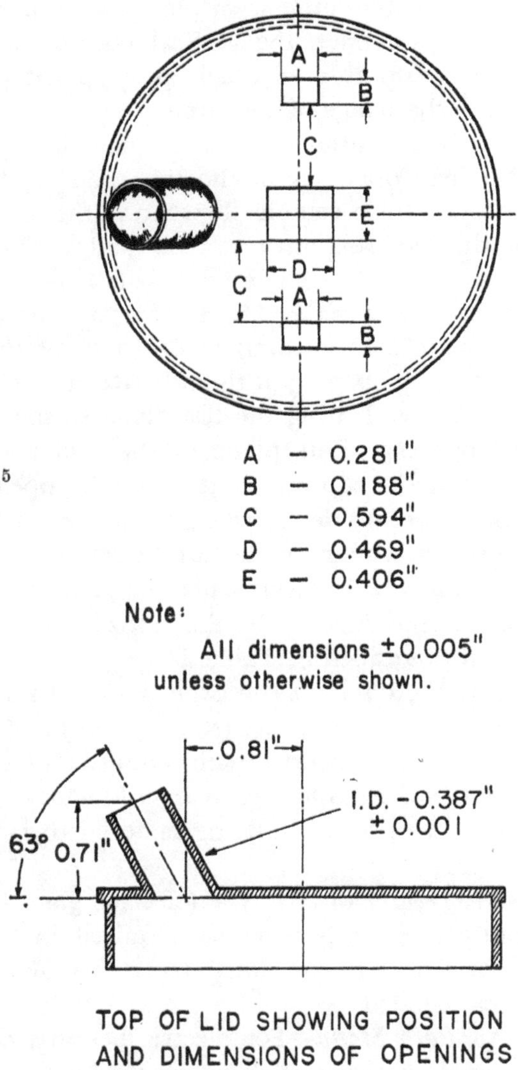

FIGURE 2. Top of Lid,[5] Showing Position and Dimensions of Openings [6]

5 All dimensions ±0.005 in. unless otherwise shown.

6 Dimensions relating to the size and position of the thermometer collar are recommended but not mandatory.

of the lid, showing the dimensions and positions of the three holes opened and closed by the shutter, and the size and position of the opening for the cup thermometer.

(*b*) The rim shall fit the collar of the liquid bath with a clearance not exceeding 0.002 in. (0.05 mm) and shall be slotted in such a manner as to press the lid firmly down on the top of the cup when the latter is in place in the bath. When this requirement is not met, the vertical position of the cup in the bath shall be suitably adjusted, as by placing a thin ring of metal under the flange of the cup.

(*c*) The shutter shall be of such size and shape that it covers the three openings in the lid when in the closed position and uncovers them completely when in the open position. The nozzle of the flame-exposure device shall conform to the dimensions given in Table 2. The device shall be designed and constructed so that opening the shutter depresses the tip to a point approximately 0.08 in. (2 mm) to the right of the horizontal center of the middle opening of the lid. (Refer to the lower part of Figure 3.) This will bring the test flame to the approximate center of the opening. The plane of the underside of the lid shall be between the top and bottom of the opening in the tip of the flame-exposure device when the latter is fully depressed.

(*d*) The collar for the cup-thermometer ferrule shall be set at an angle which permits placement of the thermometer with its bulb approximately in the horizontal center of the cup, at a depth prescribed in Table 2.

A3. *Liquid Bath* conforming to the limiting or minimum dimensions shown in Figure 3. It shall be of brass, copper, or other noncorroding metal of substantial construction. Sheet metal of about No. 20 B & S gage is satisfactory. It may, if desired, be lagged with heat-insulating material to facilitate control of temperature.

A4. *Heater* of any type (electric, gas, alcohol, etc.) capable of controlling temperature as required in Section 6. An external electric heater, controlled by a variable-voltage transformer, is recommended.

A5 *Bath Stand*—For electric heating, any type of stand may be used. For alcohol lamp or gas burner, a stand, as illustrated in Figure 1, to protect the flame from air currents (unless tests can be made in a draft-free room), is required.

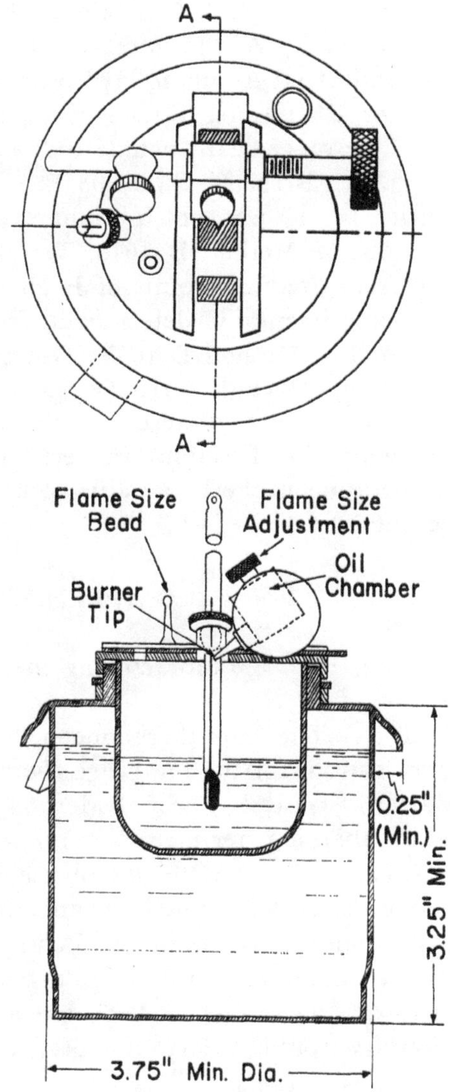

FIGURE 3. Section of Liquid Bath and Cup

APPENDIX 2

Checking Condition and Operation of Tag Closed Testers

A6. *Material*—*p*-Xylene, conforming to the following requirements:

Specific gravity (60/60°F). . . . .0.860 minimum, 0.866 maximum.

Boiling range.....2°C maximum from start to dry point, when tested by ASTM Method D 850, Test for Distillation of Industrial Aromatic Hydrocarbons, or Method D 1078, Test for Distillation Range of Lacquer Solvents and Diluents. The range shall include the boiling point of pure *p*-xylene, which is 138.35°C (281.03°F).

Purity.....95 per cent minimum calculated in accordance with ASTM Method D 1016, Test for Determination of Purity from Freezing Points of High-Purity Compounds, from the experimentally determined freezing point, measured by ASTM Method D 1015, Test for Measurement of Freezing Points of High-Purity Compounds for Evaluation of Purity.

A7. *Procedure*—Determine the flash point of the *p*-xylene, following the directions in Sections 4 to 7. When the tester is operating properly, a value of 81 ± 1°F (27.2 ± 0.6°C) will be obtained.

APPENDIX 3

Manufacturing Standardization

A8. (*a*) The cup thermometer, which conforms also to the specifications for the low-range thermometer used in the Pensky-Martens flash tester, ASTM Method D 93, is frequently supplied by the thermometer manufacturer with a metal ferrule intended to fit the collar on the lid of the flash tester. This ferrule is frequently supplemented by an adapter which is used in the larger-diameter collar of the Pensky-Martens apparatus. Differences in dimensions of these collars, which are immaterial in their effect on the results of tests, are a source of considerable unnecessary trouble to manufacturers and suppliers of instruments, as well as to users.

FLASH AND FIRE POINTS BY THE CLEVELAND OPEN CUP (MARCUSSON FLASH POINT)

The method of test recommended for determining flash and fire points by the Cleveland open cup is exactly the same as that described in the current version of ASTM D 92.

In certain countries, the use of the Marcusson cup is frequently specified for the flash point. This test is seldom required in the United States, and many laboratories do not have the necessary apparatus. Whenever a Marcusson flash point is required, it is permissible to make the determination in the Cleveland cup, adding 10°F to the observed result and reporting the sum as the flash point by the Marcusson cup.

The current issue of D 92 contains a new flash taper, as shown in Figure 1. Existing assemblies of the previous type are not to be considered obsolete and may continue to be used if desired.

ASTM DESIGNATION: D 92–56

Scope

1. This is a test procedure for determining the flash and fire points of all petroleum products except fuel oils[1] and those having an open-cup flash below 175°F (79°C).

Outline of Method

2. The test cup is filled to a specified level with sample. The

[1] This method may occasionally be specified for the determination of the fire point of a fuel oil. For the determination of the flash points of fuel oils, use ASTM Method D 93, Test for Flash Point by Means of the Pensky-Martens Closed Tester. ASTM Method D 93 should also be used when it is desired to determine the possible presence of small but significant concentrations of lower-flash-point substances which may escape detection by Method D 92.

temperature of the sample is increased fairly rapidly at first and then at a slow constant rate as the flash point is approached. At specified intervals, a small test flame is passed across the cup. The lowest temperature at which application of the test flame causes the vapors above the surface of the liquid to ignite is taken as the flash point. To determine the fire point, the test is continued until the application of the test flame causes the oil to ignite and burn for at least 5 seconds.

Apparatus

3. (*a*) *Cleveland Open Tester*—The apparatus consists of the test cup, heating plate, test-flame applicator, heater, and supports as described in detail in the Appendix. The assembled apparatus and heating plate are illustrated in Figures 1 and 2, respectively.

(*b*) *Shield*—A shield, 18 in. (46 cm) square and 24 in. (61 cm) high, open front, is recommended but not required.

(*c*) *Thermometer*[2]—An ASTM Open Flash Thermometer No. 11F (11C) having a range of 20° to 760°F (−6° to +400°C) and conforming to the requirements in ASTM Specifications E 1.

Preparation of Apparatus

4. (*a*) Support the tester on a level steady table in a draft-free room or compartment. Shield the top of the tester from strong light by any suitable means to permit ready detection of the flash point. Tests made in a laboratory hood[3] or any location where drafts occur are not reliable. During the last 30°F (17°C) rise in temperature prior to the flash point, care must be taken

---

[2] Whenever thermometers complying with ASTM requirements are not available, thermometers complying with the requirements for The Institute of Petroleum thermometer IP 28F Cleveland may be used, provided that the scale error has been found to comply with ASTM requirements, or calibration corrections are used.

[3] With some samples whose vapors or products of pyrolysis are objectionable, it is permissible to place the apparatus with shield in a hood, the draft of which is adjustable so that vapors may be withdrawn without causing air currents over the test cup during the final 100°F (56°C) rise in temperature prior to the flash point.

to avoid disturbing the vapors in the test cup by careless movements or breathing near the cup.

(*b*) Clean the cup with an appropriate solvent and remove all gums, carbon deposit, and oxide coating from the inside of the cup with fine steel wool until a bright metallic surface is prepared.

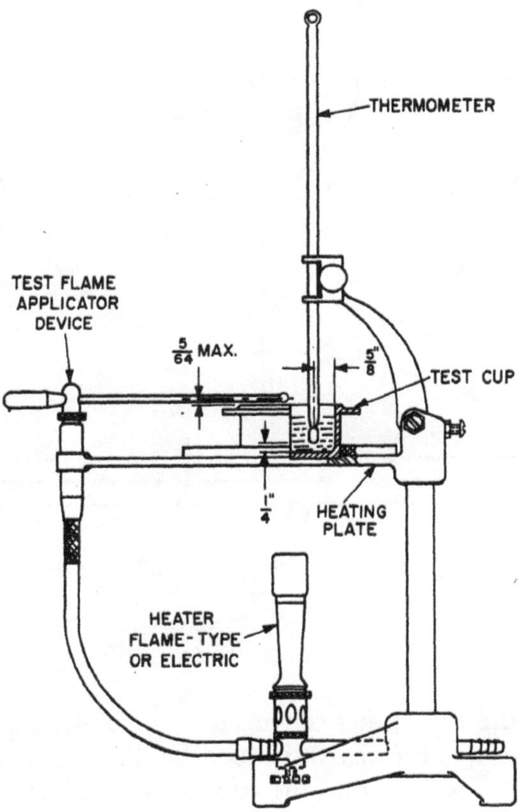

FIGURE 1. Cleveland Open Cup

(*c*) Support the thermometer[4] in vertical position with the bottom of the bulb $\frac{1}{4}$ in. (0.635 cm) from the bottom of the cup and above a point halfway between the center and back of the cup.

[4] The immersion line engraved on the thermometer will be $\frac{5}{64}$ in. (0.20 cm) below the level of the rim of the cup when the thermometer is properly positioned.

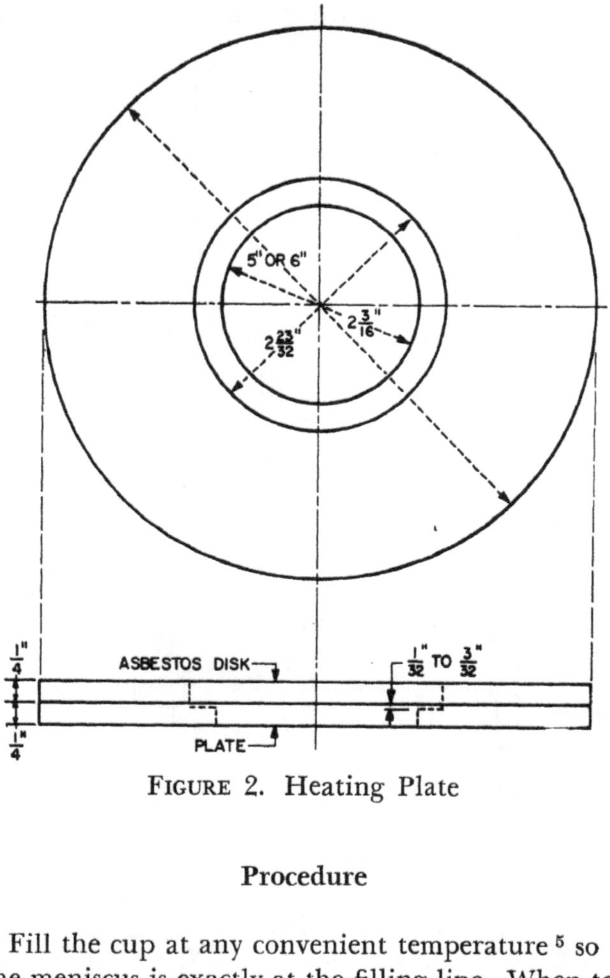

FIGURE 2. Heating Plate

Procedure

5. (*a*) Fill the cup at any convenient temperature[5] so that the top of the meniscus is exactly at the filling line. When too much sample has been poured into the cup, remove the excess, using a pipet or other suitable device; however, if there is sample on the outside of the apparatus, empty, clean, and refill it. Destroy any air bubbles on the surface of the sample.

(*b*) Light the test flame and adjust it to a diameter of $\frac{5}{32} \pm \frac{1}{32}$ in. (0.4 ± 0.08 cm), the size of the comparison bead, if one is mounted on the apparatus.

(*c*) Apply heat initially so that the rate of temperature rise

[5] Viscous samples should be heated until they are reasonably fluid before being poured into the cup; however, the temperature during heating must not exceed 100°F (56°C) below the probable flash point.

of the sample is 25° to 30°F (13.9° to 16.7°C) per minute. When the sample temperature is approximately 100°F (56°C) below the anticipated flash point, decrease the heat so that the rate of temperature rise for the last 50°F (27.8°C) before the flash point is 10° ± 1°F (5.5° ± 0.6°C) per minute.

(*d*) Starting at least 50°F (28°C) below the flash point, apply the test flame when the temperature read on the thermometer reaches each successive 5°F (2.8°C) mark. Pass the test flame across the center of the cup, at right angles to the diameter which passes through the thermometer. With a smooth, continuous motion, apply the flame either in a straight line or along the circumference of a circle having a radius of at least 6 in. (15 cm). The center of the test flame must move in a plane not more than $\frac{5}{64}$ in. (0.2 cm) above the plane of the upper edge of the cup. The time consumed in passing the test flame across the cup shall be about 1 second.

(*e*) Record as the flash point the temperature read on the thermometer when a flash appears at any point on the surface of the oil, but do not confuse the true flash with the bluish halo that sometimes surrounds the test flame.

(*f*) To determine the fire point, continue heating so that the sample temperature increases at a rate of 10° ± 1°F (5.5° ± 0.6°C) per minute. Continue the application of the test flame at 5°F (2.8°C) intervals until the oil ignites and continues to burn for at least 5 seconds. Record the temperature at this point as the fire point of the oil.

Correction for Barometric Pressure

6. If the barometric pressure at the time of the tests is less than 715 mm of mercury, record it and add the appropriate correction from Table 1 to the flash and fire points, as determined.

TABLE 1. Corrections for Barometric Pressure

Barometric Pressure, Millimeters of Mercury	Correction °F	°C
715 to 635	5	2.8
634 to 550	10	5.5

Precision

7. The following data should be used for judging the acceptability of results (95 per cent probability).

(*a*) Duplicate results by the same operator should be considered suspect if they differ by more than the following amounts:

	Repeatability
Flash Point	15°F (8.3°C)
Fire Point	10°F (5.5°C)

(*b*) The result submitted by one laboratory should not be considered suspect unless it differs from that of another laboratory by more than the following amounts:

	Reproducibility
Flash Point	30°F (16.7°C)
Fire Point	20°F (11.1°C)

APPENDIX

Apparatus

The Cleveland open tester shall consist of a test cup, heating plate, test flame applicator, heater, thermometer support, and heating-plate support, all conforming to the following requirements:

A1. *Test Cup,* conforming to the dimensional requirements shown in Figure 1 and Table 1. The cup may be equipped with a handle.

A2. *Heating Plate*—A brass, cast-iron, wrought-iron, or steel plate with a center hole surrounded by an area of plane depression, and a sheet of hard asbestos board which covers the metal plate except over the area of plane depression in which the test cup is supported. The essential dimensions of the heating plate are shown in Figure 2; however, it may be square, instead of round, and the metal plate may have suitable extensions for mounting the test-flame applicator device and the thermometer support. Also, a metal bead, as mentioned under A3, may be mounted on the plate so that it extends through and slightly above a suitable small hole in the asbestos board.

TABLE 2. Dimensional Requirements for Cleveland Open-Flash Cup

	Inches		
	Minimum	Normal	Maximum
Inside Diameter Immediately Below Filling Mark	$2\frac{15}{32}$	$2\frac{1}{2}$	$2\frac{17}{32}$
Outside Diameter Below Flange	$2\frac{21}{32}$	$2\frac{11}{16}$	$2\frac{23}{32}$
Inside Height from Center of Bottom to Rim	$1\frac{9}{32}$	$1\frac{5}{16}$	$1\frac{11}{32}$
Thickness of Bottom	$\frac{7}{64}$	$\frac{1}{8}$	$\frac{9}{64}$
Distance from Rim to Filling Mark	$\frac{23}{64}$	$\frac{3}{8}$	$\frac{25}{64}$
Distance from Lower Surface Flange to Bottom of Cup	$1\frac{7}{32}$	$1\frac{1}{4}$	$1\frac{9}{32}$
Vertical Distance from Upper Surface Flange to Rim	$\frac{7}{64}$	$\frac{1}{8}$	$\frac{9}{64}$
Thickness of Rim	$\frac{5}{64}$	$\frac{3}{32}$	$\frac{7}{64}$
Width of Lower Surface of Flange	$\frac{9}{16}$	$\frac{19}{32}$	$\frac{5}{8}$

A3. *Test-Flame Applicator*—The device for applying the flame may be of any suitable type, but it is suggested that the tip should be approximately $\frac{1}{16}$ in. (0.159 cm) in diameter at the end and the orifice, $\frac{1}{32}$ in. (0.08 cm) in diameter. The device for operating the test flame may be mounted in such a manner as to permit automatic duplication of the sweep of the test flame, the radius of swing being not less than 6 in. (15 cm) and the center of the orifice being supported so that it swings in a plane not greater than $\frac{5}{64}$ in. (0.20 cm) above the plane of the rim of the cup. It is desired that a bead, having a diameter of $\frac{5}{32} \pm \frac{1}{32}$ in. (0.4 ± 0.08 cm) is mounted in a convenient position on the apparatus so that the size of the test flame can be compared to it.

A4. *Heater*—Heat may be supplied from any convenient source. The use of a gas burner or alcohol lamp is permitted, but under no circumstances are products of combustion or free flame to be allowed to come up around the cup. An electric heater controlled by a variable-voltage transformer is preferred. The source of heat shall be centered under the opening of the heating plate, with no local superheating. Flame-type heaters may be protected from drafts or excessive radiation by any suitable type of shield that does not project above the level of the upper surface of the asbestos board.

A5. *Thermometer Support*—Any convenient device which will hold the thermometer in the specified position during a test and will permit easy removal of the thermometer from the test cup on completion of a test.

A6. *Heating-Plate Support*—Any convenient support which will hold the heating plate level and steady.

FLASH POINT BY THE PENSKY-MARTENS CLOSED TESTER

The method of test recommended for determining flash points by means of the Pensky-Martens closed tester is exactly the same as that outlined in the current version of ASTM D 93.

ASTM DESIGNATION: D 93–52

Scope

1. (*a*) This test is intended for determining the flash point of fuel oils unless the use of the Tag closed tester is specified.

(*b*) Provision is made in Sections 7 to 9 of this method for determining the flash point of cut-back asphalts and other viscous liquids and suspensions of solids.

Pensky-Martens Tester

2. The ASTM Pensky-Martens tester, a diagram of which appears in Figure 1, shall include the following major parts:

(*a*) *Cup*—The cup shall be made of brass and shall conform to the dimensional requirements prescribed in Table 1. The inside of the cup may be turned to a slightly larger diameter above the filling mark and the outside may be tapered above the flange, but the wall thickness at the upper edge shall be not less than 0.04 in. (0.102 cm). The flange should be approximately 0.5 in. (1.27 cm) in width and approximately 0.125 in. (0.318 cm) in thickness. It shall be equipped with devices for locating the position of the lid on the cup and the cup in the stove. A handle, attached permanently to the flange of the cup, is a desirable accessory. There may be a fillet approximately $\frac{5}{32}$ in. (0.397 cm) in radius inside the bottom of the cup.

(*b*) *Lid:*

(*1*) *Stirring Device*—The lid shall be equipped with a stirring device consisting of a vertical steel shaft, not less than 0.1 in. (0.254 cm) nor more than 0.125 in. (0.318 cm) in diameter,

TABLE 1. Dimensional Requirements of Pensky-Martens Flash Tester

	Inches			Centimeters		
	Minimum	Normal	Maximum	Minimum	Normal	Maximum
Inside diameter below filling mark	1.950	2.000	2.050	4.953	5.080	5.207
Difference, inside and outside diameters below filling mark	0.120	0.125	0.130	0.305	0.318	0.330
Inside height	2.150	2.200	2.250	5.461	5.588	5.715
Thickness of bottom	0.070	0.095	0.120	0.178	0.241	0.305
Distance from rim to filling mark	0.845	0.860	0.875	2.146	2.184	2.223
Distance lower surface flange to bottom of cup	1.780	1.795	1.810	4.521	4.559	4.597

mounted in the center of the cup, and carrying two two-bladed brass propellers. The blades of both propellers shall be approximately 0.313 in. (0.795 cm) wide and shall be set at an angle of approximately 45°. The smaller (upper) propeller shall have an over-all diameter of approximately 0.75 in. (1.905 cm). The larger (lower) propeller shall have an over-all diameter between 1.25 and 1.75 in. (3.175 and 4.445 cm). The thickness of the propeller blades shall be not less than 0.057 in. (0.145 cm) nor more than 0.081 in. (0.206 cm), which limits correspond respectively to No. 15 and No. 12 B. & S. gage sheet brass. The collars on which the propeller blades are mounted shall have horizontal and vertical dimensions not greater than 0.4 in. (1.016 cm).

The plane of the center of the upper propeller shall be 0.4 in. (1.016 cm) below the level of the rim of the cup. The plane of the center of the lower propeller shall be 2.0 in. (5.08 cm) below the level of the rim of the cup. The level of the rim of the cup is in effect the level of the plane part of the portion of the lower surface of the lid inside the rim.

(2) *Cover Proper*—The cover itself shall be of brass and shall have a rim projecting downward almost to the flange of the cup and fitting the outside of the cup closely. The thickness of the

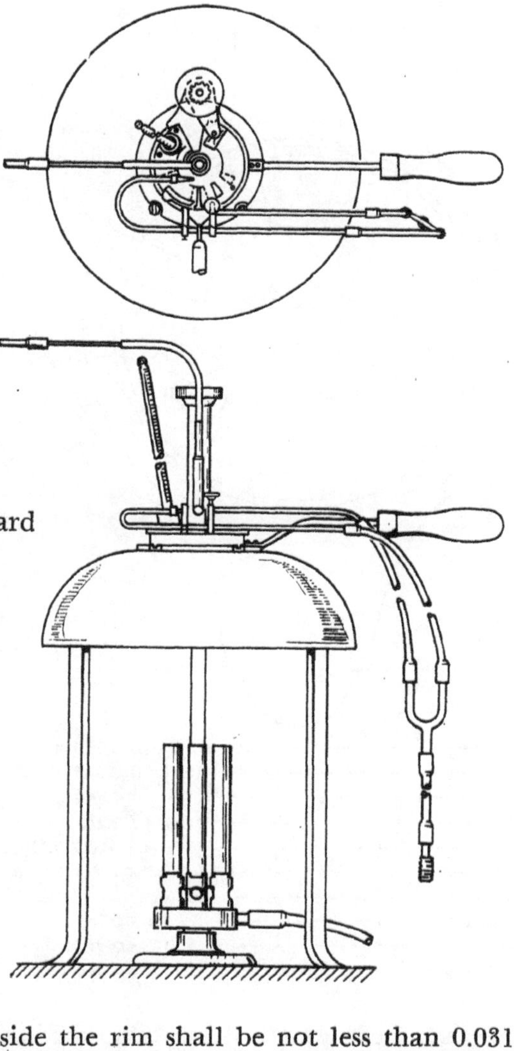

FIGURE 1. ASTM Standard Pensky-Martens Tester

cover, measured just inside the rim shall be not less than 0.031 in. (0.079 cm) nor more than 0.078 in. (0.198 cm). There shall be a proper locating device engaging with a corresponding locating device on the flange of the cup.

There shall be four openings in the cover, as indicated in Figure 2.

Opening *A* is an area defined by arcs of two concentric circles and the intersected lengths of two radii. The radius of the outer circle shall be not less than 0.938 in. (2.383 cm) nor more than 0.969 in. (2.461 cm). The radius of the inner circle shall be not

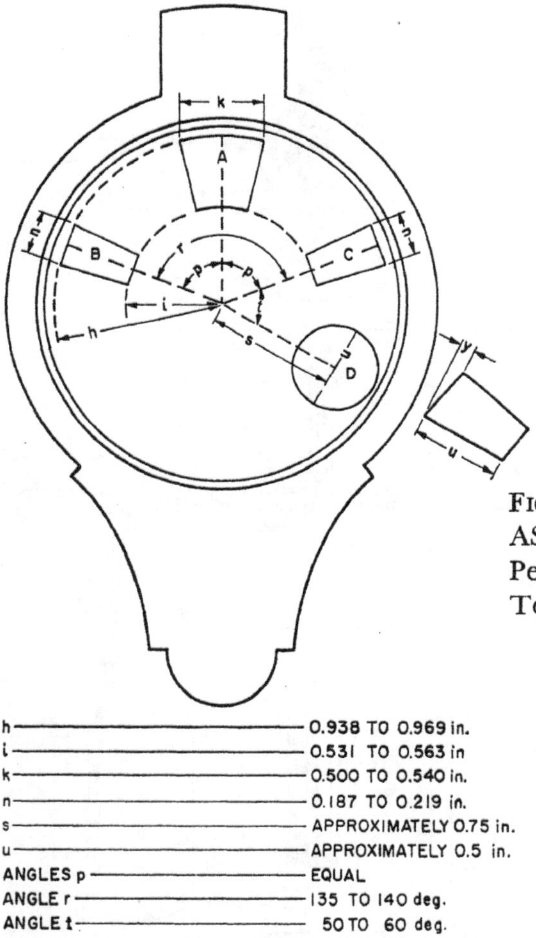

h	0.938 TO 0.969 in.
i	0.531 TO 0.563 in
k	0.500 TO 0.540 in.
n	0.187 TO 0.219 in.
s	APPROXIMATELY 0.75 in.
u	APPROXIMATELY 0.5 in.
ANGLES p	EQUAL
ANGLE r	135 TO 140 deg.
ANGLE t	50 TO 60 deg.
ANGLE y	10 TO 15 deg.

FIGURE 2. Cover for the ASTM Standard Pensky-Martens Tester

less than 0.531 in. (1.349 cm) nor more than 0.563 in. (1.430 cm). The chord of the arc of the outer circle shall be not less than 0.500 in. (1.270 cm) nor more than 0.540 in. (1.372 cm).

Openings *B* and *C* are equal areas, each of the same general form as opening *A* but of approximately half the angular width. The radii of the defining inner and outer circles shall be within the limits specified for the radii of the two circles, arcs of which partly define opening *A*. The chord of the outer arc for opening *B* or opening *C* shall be not less than 0.187 in. (0.475 cm) nor more than 0.219 in. (0.556 cm). The sum of the areas of openings *B* and *C* shall be not less than 75 per cent nor more than 100 per cent of the area of opening *A*. Openings *B* and *C* shall be equally distant from opening *A* and radii drawn through each of their centers shall be at an angle of not less than 135° nor more than 140°.

Openings *A, B,* and *C* need not conform exactly to the shape of geometrical figures bounded by arcs of two concentric circles and intersected lengths of radii. Their boundaries must, however, fall on or between the lines indicated by the limiting values of the dimensional specification of the preceding text and of Figure 2.

Opening *D* is for a thermometer collar. Its center is approximately 0.75 in. (1.905 cm) from the center of the lid and on a radius at an angle of not less than 50° nor more than 60° from a radius passing through the center of opening *C*. The thermometer collar shall have an inside diameter of approximately 0.5 in. (1.27 cm). It shall be set at an angle of not less than 10° nor more than 15° from the perpendicular.

(*3*) *Shutter*—The lid shall be equipped with a brass shutter, approximately 0.094 in. (0.239 cm) in thickness, operating on the plane of the upper surface of the lid. The shutter shall be so shaped and mounted that it rotates on the axis of the horizontal center of the lid between two stops so placed that when in one extreme position, the openings *A, B,* and *C* of the lid are completely closed and when in the other extreme position, these orifices are completely opened.

(*4*) *Flame-Exposure Device* [1]—The flame-exposure device shall

---

[1] A pilot flame for automatic relighting of the exposure flame should be provided.

A bead 0.156 in. (0.397 cm) in diameter, of some suitable material, may be mounted on the lid so that the size of the test flame can be regulated by comparison.

The mechanism operating the shutter should be of the spring type and constructed so that when at rest, the shutter shall exactly close the three openings. When operated to the other extreme, the three openings in the lid shall be exactly open and the tip of the exposure tube shall be fully depressed.

have a tip with an opening 0.027 in. (0.069 cm) to 0.031 in. (0.079 cm) in diameter. The flame-exposure device shall be equipped with an operating mechanism which, when the shutter is in the "open" position, depresses the tip so that the center of the orifice is between the planes of the under and upper surfaces of the lid proper at a point on a radius passing through the center of the larger opening *A* and approximately 0.1 in. (0.254 cm) from the outer edge of the opening.

(*c*) *Stove*—Heat shall be supplied to the cup by means of a properly designed stove which is equivalent to an air bath. This stove shall consist of (*1*) an air bath and (*2*) a top plate on which the flange of the cup rests.

(*1*) *Air Bath*—The air bath shall have a cylindrical interior 1.625 in. (4.128 cm) to 1.656 in. (4.206 cm) in depth and a diameter not less than 0.125 in. (0.317 cm) nor more than 0.156 in. (0.396 cm) greater than the outside diameter of the cup. The air bath may be either a flame-heated metal casting[2] or an electric-resistance element.[3]

(*2*) *Top Plate*—The top plate shall be of metal. The total distance from the upper surface of the plate to the bottom of the air bath shall exceed the distance from the under surface of the flange to the bottom of the cup by not less than 0.063 in. (0.160 cm) nor more than 0.125 in. (0.317 cm).

The top plate shall be mounted with an air gap between it and the air bath. The top plate may be attached to the air bath by means of three screws and spacing bushings. The spacing bushings should be of proper thickness to define the air gap which shall be not less than 0.125 in. (0.317 cm) nor more than 0.187 in. (0.475 cm). The spacing bushings shall be not more than 0.375 in. (0.952 cm) in diameter.

---

[2] If the heating element is a flame-heated metal casting, it shall be so designed and used that the temperature of bottom and walls is approximately the same. On this account, it should be not less than 0.25 in. (0.635 cm) in thickness. The casting shall be designed so that products of combustion of the flame cannot pass up and in contact with the cup.

[3] If the air bath is of the electric-resistance type, it shall be constructed so that all parts of the interior surface are heated equally. This necessitates an even distribution of resistance wire over bottom and walls and a method of construction such that heat is given out from the whole core of the resistance element rather than directly from the wire.

(*d*) *Thermometers*[4]—Two standard thermometers shall be used with the ASTM Pensky-Martens tester.

For tests in which the indicated reading falls within the limits 20° to 200°F, a low-range ASTM P.M. and Tag thermometer, graduated in either Centigrade or Fahrenheit degrees as specified, having a range of −7° to +110°C or +20° to +230°F, and conforming to the requirements for thermometers 9C or 9F, shall be used.

For tests in which the indicated reading falls within the limits 230° to 700°F, a high-range ASTM P.M. high thermometer, graduated in either Centigrade or Fahrenheit degrees as specified, having a range of 90° to 370°C or 200° to 700°F, and conforming to the requirements for thermometers 10C or 10F, shall be used.

For the range 200° to 230°F, either thermometer may be employed, depending on the convenience of the operator.

Thermometers shall be mounted so that the bottom of the bulb is 1.75 in. (4.445 cm) below the level of the rim of the cup (which corresponds to the level of the lower surface of the portion of the lid inside the rim).

Procedure

3. (*a*) All parts of the cup and its accessories shall be thoroughly cleaned and dried before the test is started. Particular care shall be taken to avoid the presence of any gasoline or naphtha used to clean the apparatus after a previous test.

(*b*) The cup shall be filled with the oil to be tested up to the level indicated by the filling mark.

(*c*) The lid shall be placed on the cup and the cup set in the stove. Care shall be taken to have the locating devices properly engaged. The thermometer shall be inserted. If it is known that the oil will flash above 220°F, the ASTM P.M. high thermometer

---

4 Whenever thermometers complying with ASTM requirements are not available, thermometers compying with the requirements for The Institute of Petroleum thermometer IP 15 F (or 15 C) P.M. low, or IP 16 F (or 16 C) P.M. high, may be used, provided calibration corrections are used in temperature ranges where the IP requirements for scale accuracy are less stringent than those of ASTM.

may be selected; otherwise, it is preferable to start with the ASTM P.M. and Tag thermometer and change in case a temperature of 220° to 230°F is reached.

(*d*) The test flame shall be lighted and adjusted so that it is of the size of a bead $\frac{5}{32}$ in. (3.97 mm) in diameter.

(*e*) Heat shall be supplied at such a rate that the temperature read on the thermometer increases not less than 9° nor more than 11°F per minute. The stirrer shall be turned at a rate of 1 to 2 revolutions per second.

(*f*) Application of the test flame shall be made at each temperature reading which is a multiple of 2°F up to 220°F. For the temperature range above 220°F, application shall be made at each temperature reading which is a multiple of 5°F. The first application of the test flame shall be made at a temperature at least 30°F below the actual flash point. Application of the test flame shall be made by operating the device controlling the shutter and test-flame burner so that the flame is lowered in $\frac{1}{2}$ second, left in its lowered position for 1 second, and quickly raised to its high position. Stirring shall be discontinued during the application of the test flame.

Flash Point

4. The flash point is recorded as the temperature read on the thermometer at the time of the flame application that causes a distinct flash in the interior of the cup. The true flash must not be confused with the bluish halo that sometimes surrounds the test flame for the applications preceding the one that causes the actual flash.

Barometric Pressure

5. The barometric pressure shall be observed and recorded. No corrections shall be made except in case of dispute when the flash point figure shall be corrected according to the following rule:

For each inch (25 mm) below 29.92 in. (760 mm) barometric reading, add 1.6°F to the flash point.

For each inch (25 mm) above 29.92 in. (760 mm) barometric reading, subtract 1.6°F from the flash point.

Precision

6. Results shall not differ from each other by more than listed in Table 2.

TABLE 2. Precision of Flash Point

Flash Point, °F	Repeatability, One Operator and Apparatus, °F	Reproducibility, Different Operators and Apparatus, °F
200 and under	2	4
Over 200	5	10

DETERMINATION OF THE FLASH POINT OF CUT-BACK ASPHALTS AND OTHER VISCOUS LIQUIDS AND SUSPENSIONS OF SOLIDS

Apparatus

7. The apparatus shall consist of the following:

(*a*) *Flash Point Tester*—Pensky-Martens tester as described in Section 2 (*a*), (*b*), and (*c*), except that the stirrer shall be mechanically operated to stir in a downward direction at a speed of 70 to 80 rpm.

(*b*) *Thermometer*—Low-range thermometer as specified in Section 2 (*d*).

Procedure

8. The sample to be tested and the tester shall be brought to a temperature 30°F lower than the estimated flash point. The air space between the cup and the interior of the air bath shall be completely filled with water at the temperature of the tester and sample. The temperature shall be raised[5] throughout the

[5] Solid carbon dioxide (dry ice) shall, in no case, be used to obtain the proper rate of temperature rise, since CO_2 has a blanketing effect which leads to a false flash point.

duration of the test at a rate of not less than 2° nor more than 3°F per minute. With the exception of this requirement for rate of heating and the rate of stirring of 70 to 80 rpm, specified in Section 7 (*a*), the procedure shall be the same as that prescribed in the standard method of test for flash point by Tag closed tester (ASTM Designation: D 56).

9. Duplicate tests should not differ by more than 5°F.

DETERMINING THE NONVOLATILE MATTER IN CTLA POLYMERS

This method is essentially the same as described in ASTM D–154 for the examination of varnishes. It is intended primarily for testing the various types of CTLA Polymer, but may be used also on other products where specified.

Apparatus

(1) *Evaporating Dish*—A Petri dish partly filled with clean, dry sand should be used. The dish should measure approximately 9 cm in diameter and 1 cm in depth. A satisfactory dish may be obtained from the Scientific Glass Apparatus Company, Bloomfield, N.J., under their catalog number 7170; it is available only in pairs.

(2) *Weighing Bottle*—A weighing pipette or a small stoppered bottle fitted with a dropper will be needed.

Procedure

(1) Approximately 30 g of clean, dry sand should be placed in a Petri dish and spread evenly over the surface. The dish should then be dried in an oven at 105° to 110°C for 2 hours, cooled in a desiccator, and weighed to the nearest 0.1 mg.

(2) The weighing bottle or pipette should be cleaned thoroughly, dried in an oven, and cooled in a desiccator. It should then be partly filled with the sample to be tested and weighed to the nearest 0.1 mg. Approximately 1.5 g of the sample should be transferred to the Petri dish and the bottle again weighed, recording the difference as the weight of the sample. During the transfer, the liquid should be dropped uniformly over the sand.

(3) The dish should be heated for 3 hours in an oven maintained at 105° to 110°C. At the end of the 3 hours, it should be removed, cooled in a desiccator, and reweighed. The increase over the original weight of the dish should be calculated to percentage and reported as *nonvolatile matter.*

L.I.C. No. 10.91

RAPID METHOD FOR COMPARATIVE EVAPORATION RATES OF SOLVENTS

The evaporation rate of a solvent intended for use in the paint and lacquer industry is an important consideration, since it will be a major factor in the drying time of the finished product. By the following method, this information can be obtained in a few minutes. The procedure was developed by Subcommittee 66 of the New York Paint and Varnish Production Club and provides a means for comparing the evaporation rates of different solvents.

The method is applicable to solvents used in the paint and lacquer industry, which normally consist of hydrocarbons, alcohols, ketones, and/or mixtures of organic esters.

Some discrepancies have been noted for a blend of octane and xylene.

The *evaporometer* consists of an ordinary cone-shaped centrifuge tube containing a side arm near the 25-ml mark. A measured amount of the sample is placed in the tube, which is then mounted in a constant-temperature bath in the horizontal plane. A dry gas, either air or nitrogen, is forced into the mouth of the tube at a constant rate and is vented through the side arm. At periodic intervals, the tube is removed from the bath, held in the vertical position, and the amount of liquid remaining in it noted. These observations are repeated until the liquid is evaporated almost to dryness, and the observed data plotted on a graph.

The assembly is standardized by evaporating a standard of known composition and adjusting conditions until 97.5 volume per cent evaporates within a given time. The unknown is then evaporated under the same conditions. In general, the recommended standard is 2° toluene, although this is subject to agreement between buyer and seller.

Apparatus

(1) *Centrifuge Tube* as shown in Figure 1, and consisting of an ordinary cone-shaped tube with a side arm near the 25-ml mark. It is imperative that the marks below 2 ml are accurate.

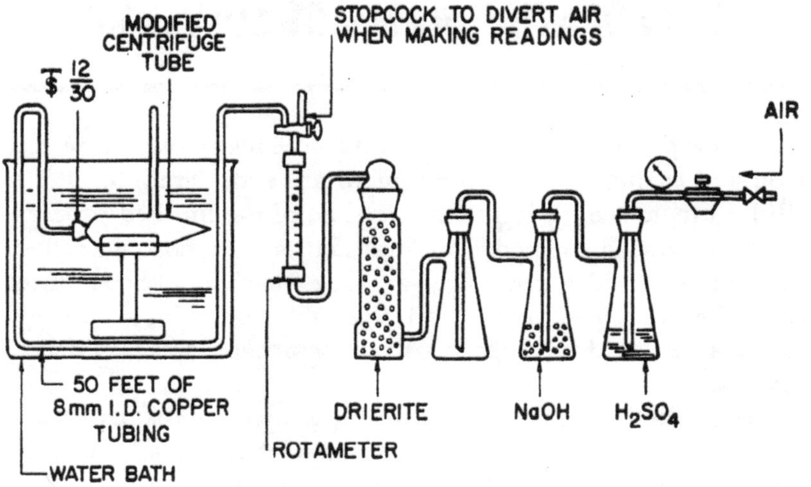

FIGURE 1. Modified Wetlaufer-Gregor Apparatus

Provision must be made for delivering a dry gas to the mouth of the tube, and the exact location of the gas inlet tube is important. If the centrifuge and air-delivery tubes have ground-glass joints, this location will be automatic. If a cork stopper is used, the bottom of the stopper should be flush with the shoulder of the centrifuge tube, and the end of the gas-inlet tube, flush with the end of the stopper.

(2) *Constant-Temperature Bath* that can be maintained at a temperature of 25° ± 0.1°C and is fitted with the following accessories:

(a) Some means of holding the tube in the horizontal position below the surface of the liquid. A convenient arrangement consists of a clamp for the tube which is attached to a bar across the top of the bath by a swivel joint.

(b) A 50-ft length of 8-mm internal diameter copper tubing for preheating the gas to be delivered to the tube.

The copper tubing may be coiled in any convenient manner, either in a few large loops at the bottom of the bath or in a spiral near one side.

(c) A thermometer covering the temperature of 25°C, and readable to at least 0.2°. The Saybolt viscosity type is satisfactory.

(3) *Rotameter* covering the range of 0 to 25 liters per minute. It is available from the Brooks Rotameter Co., Lansdale, Pa., under their designation Tube 6-15-2. The Rotameter may be equipped with floats of different composition so that the rate of flow can be adjusted within wide limits. For this test, the stainless-steel float should be used. Provision must be made for interrupting the flow of gas to the evaporometer at periodic intervals. This may be done either by attaching a three-way stopcock to the rotameter, as shown in Figure 1, or by having a valve located in the delivery line.

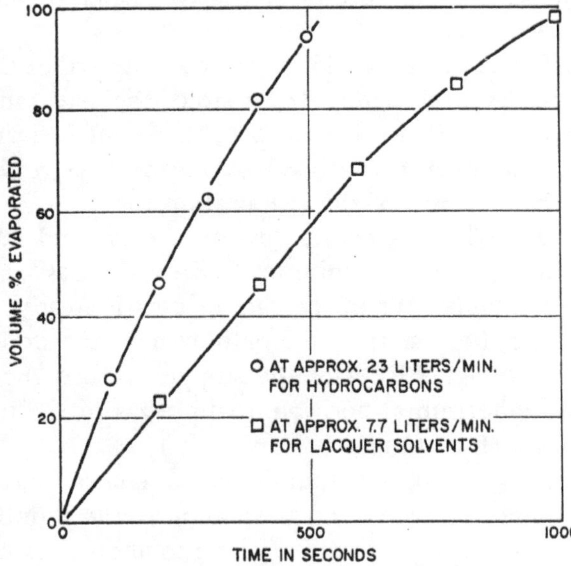

FIGURE 2. Evaporation Rate of Solvesso Toluol Standard at 25°C

The gas which is used must be dry. If compressed air is used, the complete assembly must include the drying towers shown in Figure 1; if cylinder nitrogen is used, these towers may be omitted. In any case, provision must be made for controlling the flow of gas through the assembly and measuring the pressure in it. The assembly should, therefore, include a manometer, which is attached to the gas line between the copper tubing and the evaporation tube.

(4) *Stopwatch* or electrical timer graduated in seconds.

Calibration

(1) *Calibration Standard*—The usual standard is 2° toluene.

(2) Fill the bath with water and adjust its temperature to 25° ± 0.1°C. Insert the coil of copper tubing, and attach the discharge end to one arm of a T-joint. Attach the second opening of the T to the manometer, and the third, to the centrifuge tube. Mount this tube in the bath so that it will be completely immersed in the liquid, but the opening of the side arm will be above the liquid level. Connect the other end of the copper tubing to the top of the rotameter, which is mounted in the vertical position. Attach the bottom of the rotameter to the source of gas.

Some provision must be made for the safe disposal of the fumes from the samples under test. If possible, the bath should be mounted in a hood. If this is not possible, mount it near a hood and void the fumes into the hood by means of a rubber tube attached to the side arm of the evaporation tube.

(3) When everything is ready, turn on the gas and adjust the flow to about 23 liters per minute, leading the gas completely through the assembly. When the rate of flow is steady, stop the flow of gas, raise the tube from the bath, remove the stopper, and add 2.0 ml of toluene. Replace the stopper, return the tube to the bath in the horizontal position, reapply gas to the tube, and start the stopwatch or timer.

It is important that the sample never comes into contact with the stopper or the ground-glass joint at the mouth of the tube. Changing the tube from one plane to another must, therefore, be done with great care.

(4) At periodic intervals, stop the flow of the gas and the stopwatch. Record the reading of the stopwatch, remove the tube from the bath, allow it to drain for at least 1 minute, and note the volume of liquid remaining in it. Return the tube to the bath, turn on the gas, and restart the stopwatch. Continue

these observations until 97.5 per cent of the sample has evaporated.

During these observations the following are important:

(a) There must be a total of five observations, no more and no less.

(b) The drainage time is not included in the over-all evaporation time. The watch is stopped as soon as the flow of gas is stopped and is started again, without returning to the zero point, when the gas is reapplied.

(c) The test is stopped when the volume of liquid in the tube has been reduced to 0.05 ml (i.e., 97.5 per cent has evaporated).

(5) When the test is stopped, the reading on the timer should be 520 ± 9 seconds. If it is not, repeat the calibration, changing the rate of flow of the gas as necessary. On subsequent observations, make the inspections at the proper time intervals so that the amount of liquid which has been evaporated during any single period will be about 20 per cent of the total. Also note and record the readings of the manometer.

(6) When the total time is within the permissible limits, prepare a graph similar to the one shown in Figure 2, plotting the percentage of evaporated against elapsed time. This curve should be even, without pronounced peaks or valleys.

(7) When testing conditions have been established, repeat the evaporation once more and compare the observed results. At any given point, individual readings should not differ from the mean by more than 3 per cent of the mean. If this is true, the assembly is ready for use; if it is not true, repeat as necessary until concordant results are obtained.

A small variation in the rate of flow of the gas may affect the final results materially. The manometer is used largely as a means of measuring accurately the rate of flow. The readings should be noted at frequent intervals, and adjustments made as necessary. It is important that the evaporation time is as close to 520 as possible. The limit of ±9 seconds is simply an arbitrary figure; actually, the more closely the operator can approach 520, the more accurate the results on a sample will be.

(8) The listed testing conditions are used when the sample under test is a simple solvent, such as a hydrocarbon. When

the sample is a lacquer solvent or a mixture of compounds which will evaporate quickly, the testing conditions must be adjusted correspondingly. All details of procedure are as outlined, with the single exception that the rate of flow of the gas is adjusted so that the time required for the evaporation of the standard will be 985 ± 17 seconds. To do this, the flow of gas must be adjusted to about 7.7 liters per minute.

Note Carefully: The standards are subject to agreement between buyer and seller. It is conceivable that either party might specify an entirely different standard, other testing conditions, etc. In that case, the conditions should be modified according to any agreement which is satisfactory to, and understood by, both parties.

Procedure

(1) After the assembly has been calibrated, repeat the entire procedure with the sample under test, keeping the rate of flow of the gas the same as during the calibration. Make a trial run to determine the approximate time for 97.5 volume per cent to evaporate. Note carefully the amounts evaporated at the time of of each inspection.

(2) After the trial test has been made, repeat the test, this time paying particular attention ot the timing of the inspections. As nearly as possible, make each inspection as soon as 20 per cent of the sample has been evaporated.

Being a single compound, toluene presents no particular problem, and the evaporation curve should be even and regular. However, if the sample is a mixture containing appreciable quantities of one or more volatile liquids, mixed with heavier ones, the evaporation losses will be high during the first minutes and will taper off afterward. Inspections should, therefore, be made at intervals of the amount evaporated, rather than of the time.

(3) Repeat the evaporation, and compare the results. At any single point on the curve, neither reading should differ from the mean by more than 3 per cent of the mean. If necessary, repeat until concordant results are obtained.

Report the percentage evaporated at various time intervals, Depending on the way the specification is written, the report may be worded:

100 seconds, ... per cent evaporated (at .... liters/minute), or 20 per cent evaporated .... seconds (at .... liters/minute)

In some cases, the specifications may require that the sample is equal to a certain standard within prescribed limits. If the observed results are within the limits, report as "Equal to standard" or "Passes."

It is vitally important that the centrifuge tube is clean at the start of the test. Before a test is started, the tube should be washed with soap and water, soaked with chromic acid, rinsed with tap and distilled water until acid free, and dried in an oven. Between tests on the same sample, the tube may be rinsed with acetone and dried by a stream of clean warm air.

GRAVITY OF PETROLEUM PRODUCTS BY THE HYDROMETER

The test method recommended for determining the gravity of petroleum products by means of the hydrometer is exactly the same as that described in the current version of ASTM D 287, subject to the qualifications outlined in the following paragraphs.

There are two applicable methods in the D-2 Manual for determining gravity by hydrometer. One of these is the familiar D 287, and the results are reported in terms of API degrees. The other is numbered D 1298, and the results are reported in terms of specific gravity at 60/60°F. Since the only essential differences in the two methods are the use of the proper hydrometer and the terms of the report, the following method may be considered as the equivalent of both D 287 and D 1298.

In D 287, various types of hydrometers are listed as permissible alternates. The plain long form, as listed under ASTM numbers 1 to 10H inclusive, should be used wherever possible, and is mandatory for referee or precision testing. The other types should be used only when the amount of sample is limited, or when the use of the thermohydrometer is definitely required by special instructions.

Under certain conditions, the sample may be tested at temperatures which are materially different from room temperature, as outlined in Table 2 of both D 287 and D 1298. Whenever this is done, particular attention must be paid to Section 5 (f) of the ASTM method. In other words, the temperature of the sample before and after the test must not differ by more than 1°F. This is particularly important when testing such products as waxy lubricants or fuels, microcrystalline waxes, and samples of CTLA Polymer, aromatic tars, etc., which are solid or semisolid at room temperature.

ASTM DESIGNATION: D 287–55

Scope

1. This is a procedure for the determination by means of a glass hydrometer of the API gravity (in vacuum) of crude petroleum and of petroleum products normally handled as liquids and having a Reid vapor pressure of 26 lb or less. Results are determined at 60°F, or converted to values at 60°F, by means of standard tables.

Definition

2. *API Gravity* is defined by the following equation:

$$\text{API Gravity, degrees} = \frac{141.5}{\text{sp. gr. } 60/60°\text{F}} - 131.5$$

Apparatus

3. The following apparatus is required:

(*a*) *Hydrometers,* of glass, graduated in API degrees (in vacuum) as listed in Table 1. For referee testing of petroleum products, the long plain form of hydrometer (1H to 10H) should be used

TABLE 1. Available API Hydrometers

ASTM Designation	Type	API Range Total	API Range Each Hydrometer
1H to 10H	long plain form	−1 to 101	12
21H to 40H	short plain form	0 to 101	6
51H to 60H	thermohydrometer	−1 to 101	12
71H to 74H	thermohydrometer	−1 to 41	12

(*b*) *Thermometers,* having a range of −5° to +215°F.

(*c*) *Hydrometer Cylinders,* made of metal, clear glass, or plastic. For convenience in pouring, the cylinder may have a lip on the rim. The inside diameter of the cylinder shall be at least 25 mm greater than the outside diameter of the hydrometer used in it. The height of the cylinder shall be such that the length of the column of sample it contains is greater by at least 25 mm than the portion of the hydrometer which is immersed beneath the surface of the sample.

Temperature of Test

4. The gravity determined by the hydrometer method is most accurate at or near the standard temperature of 60°F; use this or any other temperature between 0° and 195°F for the test so far as it is consistent with the type of sample and necessary limiting conditions shown in Table 2.

Procedure

5. (*a*) Adjust the temperature of the sample in accordance with Table 2. The hydrometer cylinder and thermometer shall be at approximately the same temperature as the sample to be tested.

(*b*) Pour the sample into the clean hydrometer jar without splashing, so as to avoid the formation of air bubbles and to reduce to a minimum the evaporatoin of the lower-boiling constituents of the more volatile samples. For the more volatile samples, transfer to the hydrometer cylinder by siphoning. Remove any air bubbles formed, after they have collected on the surface of the sample, by touching them with a piece of clean filter paper before inserting the hydrometer. Place the cylinder containing the sample in a vertical position in a location free from air currents. Take precautions to prevent the temperature of the sample from changing appreciably during the time necessary to complete the test. During this period, the temperature of the surrounding medium should not change by more than 5°F.

(*c*) Lower the hydrometer gently into the sample and, when it has settled, depress it about two scale divisions into the liquid and then release; keep the rest of the stem dry, as unnecessary liquid on the stem changes the effective weight of the instrument and so affects the reading obtained. With samples of low viscosity, a slight spin imparted to the instrument on releasing assists in

TABLE 2. Limiting Conditions and Test Temperatures

Sample Type	Gravity Limits	Initial Boiling Point Limits	Other Limits	Test Temperature
Highly volatile	Lighter than 70° API	...	...	Cool to 35°F or lower in original closed container
Moderately volatile	Heavier than 70° API	Below 250°F	...	Cool to 65°F or lower in original closed container
Moderately volatile and viscous	Heavier than 70° API	Below 250°F	Viscosity too high at 65°F	Heat to minimum temperature for sufficient fluidity
Nonvolatile	Heavier than 70° API	Above 250°F	...	Any temperature between 0° and 195°F as convenient
Mixtures of non-petroleum with petroleum products	...	...	...	60 ± 0.25°F

bringing it to rest, floating freely away from the walls of the hydrometer cylinder. Allow sufficient time for the hydrometer to become completely stationary and for all air bubbles to come to the surface. This is particularly necessary in the case of the more viscous samples.

(*d*) When the hydrometer has come to rest, floating freely, and the temperature of the sample is constant to 0.2°F, read the hydrometer to the nearest scale division. The correct reading is that point on the hydrometer scale at which the surface of the liquid cuts the scale. Determine this point by directing the eye slightly below the level of the liquid and slowly raising it until the surface, first seen as a distorted ellipse, appears to become a straight line cutting the hydrometer scale.

(*e*) To make a reading with nontransparent oils, observe the point on the hydrometer scale to which the sample rises above its main surface, directing the eye slightly above the plane surface of the liquid. This reading requires a correction. Determine this correction for the particular hydrometer in use by observing the height above the main surface of the liquid to which the oil rises on the hydrometer scale when the hydrometer in question is immersed in a transparent oil having a surface tension similar to that of the sample under test.

(*f*) Observe the temperature of the sample to the nearest 0.25°F immediately before and after the observation of the gravity, the liquid in the cylinder being thoroughly but cautiously stirred with the thermometer, the whole of the mercury thread being immersed. Should these temperature readings differ by more than 1°F, repeat the temperature and gravity observations when the temperature of the sample has become more stable. Record the mean of the thermometer reading before and after the final hydrometer reading, to the nearest Fahrenheit degree, as the temperature of the test.[1]

[1] When thermohydrometers are used, stir the sample by carefully raising and lowering the hydrometer. It is satisfactory in this case to read the thermometer scale after the hydrometer reading has been observed. Because the thermometer incorporated in the hydrometer has certain inherent defects of design, the precision of the reported value will be poorer than that described in Section 7.

Calculation

6. (*a*) When gravities have been observed on opaque liquids by the procedure given in Section 5(*e*), subtract the correction in API degrees from the hydrometer reading observed.

(*b*) Correct all hydrometer readings to 60°F, using Table 5, "Reduction of Observed API Gravity to API Gravity at 60°F," of the *ASTM-IP Petroleum Measurement Tables* (American Edition).[2]

7. The criteria in Table 3 should be used for judging the acceptability of results obtained at temperatures differing from 60°F by less than 18°F:

TABLE 3. Precision of Hydrometer Measurements

	Repeatability Duplicate Results by the Same Operator	Reproducibility Average of Two Results in Each of Two Laboratories
API Degree	0.2	0.5

[2] Published jointly by, and available from, the American Society for Testing Materials, 1916 Race St., Philadelphia 3, and the Institute of Petroleum, 26 Portland Place, London W-1.

SPECIFIC GRAVITY BY MEANS OF THE PYCNOMETER

This method is applicable to any type of liquid or solid petroleum product, except certain specialties. Report the results in terms of specific gravity or density.

Specific gravity is the ratio of the weights of equal volumes of the substance under test and distilled water. Temperature, of course, plays an important part in this ratio, and is included in the final report. A typical specific gravity report reads: "Sp. Gr. 0.753 20°/20°C." This means that a given volume of the sample at 20°C will weigh 0.753 times as much as the same volume of distilled water at the same temperature. Occasionally specific gravities are determined with the sample at one temperature, and the water at another. As an example, a report reading "90°/60°F" would mean that the weight of a given volume of the sample at 90°F was compared with the weight of the same volume of water at 60°F.

In most petroleum work, the reference temperature is 60°F for both the water and the sample, and is so understood if no temperature is mentioned in the report. If the temperature of either the sample or the standard is other than 60°F, this fact must be clearly indicated. Specific gravities of solvents, such as alcohol, ketone, etc., are frequently taken on a basis of either 20°/20°C or 25°/25°C; in all such cases, the temperature must be reported.

Density is the actual weight, in grams, of a unit volume of the sample under test, and is commonly reported either as absolute or as relative. Absolute density is the weight in grams of 1 cc of substance, while relative density is the weight in grams of 1 ml. In this section, results are determined in, and reported as, units of relative density.

Apparatus

The general types of pycnometer are shown in Figure 1. The first is the most common and is preferred for nonviscous products where the amount of sample is insufficient for an estimation by the hydrometer. The second is preferred for various organic solvents, such as alcohol, ketone, etc. The others may be used for viscous, semisolid, or solid products, which will not flow readily through the small neck of the normal pycnometer. Any of these types may be obtained from the regular supply houses.

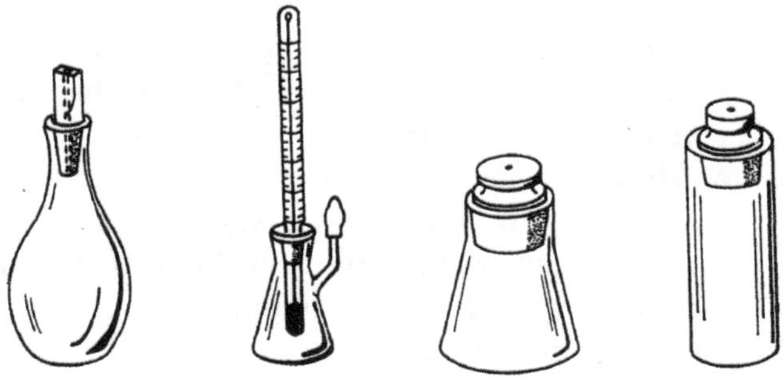

FIGURE 1. Pycnometers

Calibration

(1) Whatever the type selected, clean the pycnometer thoroughly, dry it for 1 hour at 220°F, and store in a desiccator. When cool, weigh it to the nearest 0.1 mg.

(2) Fill the tared pycnometer to overflowing with distilled water which has recently been boiled and then cooled to below the test temperature. Insert the stopper, partly immerse the filled pycnometer in a water bath maintained at the desired temperature, and leave it undisturbed for at least fifteen minutes.

(3) When the contents have come to bath temperature, remove the pycnometer and without delay wipe off any excess water from the *top* of the stopper. Wipe the rest of the outside with a clean dry cloth, taking care not to disturb the small drop of water which may form on the top of the stopper. When dry, weigh to the nearest 0.1 mg and record the increase in weight.

(4) When specific gravity is being determined, the increase in weight, in grams, is recorded as the weight of the "equal volume of water."

(5) When density is being determined, find from proper tables the density of water at the test temperature. Divide the increase

in weight, in grams, by the density and record the quotient as the volume of the bottle at that temperature.

Suggested values are:

Temperature °C	Density
4.00	1.00000
15.00	0.99913
15.56 (60°F)	0.99904
20.00	0.99823
25.00	0.99707

Procedure—Specific Gravity

Nonviscous Liquids

(1) The procedure is the same as for the calibration, the only difference being that the pycnometer is filled with the sample instead of water.

(2) Divide the net weight of the sample in the pycnometer by the weight of the equal volume of water and report the quotient as the specific gravity of the sample under test.

(3) If, for any reason, the temperature of either the sample or the water is other than 60°F, include this information in the report.

Viscous Liquids

(1) Heat the sample to a temperature at which it will pour freely and hold it at that temperature for about 15 minutes. Fill the tared pycnometer about three-quarters full, taking care that none of the sample touches the ground portion of the neck. If any of the sample should touch this portion, wipe it off immediately with a clean dry cloth. Insert the stopper, put the pycnometer into a vacuum desiccator, and allow it to cool to room temperature.

During the filling and heating it is vitally important that no air bubbles are included in the molten product.

(2) When cool, weigh the pycnometer to the nearest 0.1 mg and record the net weight of sample.

(3) Remove the stopper, fill the pycnometer to overflowing with distilled water, and complete the test as described under calibration. Note and record the weight, in grams, of added water.

(4) Calculate the specific gravity of the sample by the equation:

$$G = \frac{S}{V - W}$$

where

- S is the weight of sample in the pycnometer, in grams.
- V is the weight in grams of water in the pycnometer as obtained under CALIBRATION (4).
- W is the weight of added water.
- G is the specific gravity of the sample under test.

Procedure—Density

(1) Calculate the volume of the pycnometer as described under CALIBRATION (5).

(2) From here on the procedure is the same as previously described, the only difference being that in the final calculations the values V and W are expressed in units of volume, rather than weight. In each case, the weight of water in the pycnometer must be divided by the relative density of water at test temperature to obtain the proper volume.

This is a rapid routine procedure for plant operations. Since the average analytical balance has a *theoretical* accuracy of 0.0001 g, the results obtainable with this method are probably accurate to only about 0.0002 unit. If greater accuracy is wanted, other types of apparatus must be used and corrections must be made to reduce all weighings to a vacuum basis. Details are outlined in ASTM D 1217.

VISCOSITY DETERMINATION BY THE SAYBOLT VISCOMETER

The method recommended for determining viscosity by the Saybolt viscometer, either Universal or Furol, is exactly the same as that described in the current version of ASTM D 88, subject to the qualifications outlined in the following paragraphs:

(1) In the present version, some accessory items, such as the strainer and withdrawal pipette, have been changed. The similar items listed in previous versions need not be discarded.

(2) D-88 recommends the use of certain standard oils available through the API, or other sources. While these may be used, a number of satisfactory standards are available from much cheaper sources. These may be substituted if desired.

(3) Under certain conditions, the regular routine may be modified slightly. The modified method is the same as in ASTM D 88 in all respects, with the single exception that the receiver has a capacity of 28.6 ml, rather than the usual 60. If the observed result is multiplied by 3, the product will be the same as that obtainable by the regular method. This modification is useful when the sample under test has an excessively long running time and contains considerable quantities of suspended matter.

It is strongly recommended that Saybolt viscosity tubes made of stainless steel should be used, which, with proper care, should last indefinitely. The ASTM method specifies that the tubes shall be made "entirely of corrosion-resistant metal," with no statement as to composition. Tubes made of brass would satisfy this requirement, but could easily be damaged in service.

The complete assembly may be ordered for either Saybolt Universal or Furol, as desired. It will consist of a tip and a barrel and gallery, each unit being stamped with a serial number. This combination of barrel and tip MUST NOT BE CHANGED IN ANY MANNER. In particular, the Universal tip should not be removed and replaced with the Furol, and vice versa.

A number of standard-viscosity oils are available. Since no oil will reliably hold a given viscosity value for an indefinite period, the standard sample should be purchased in small amounts at frequent intervals. It is strongly recommended that standard-viscosity oils are kept in each laboratory for a periodic check of the instruments in use. The Saybolt, Redwood, and kinematic values should be those actually obtained in master instruments of the proper type, and not values converted from one system to another. The kinematic values will be based on the value for water which became effective as of July 1, 1953.

In earlier years, it was recommended that the viscosity of light oils should be determined by the Saybolt 115 viscosimeter. This instrument is now obsolete. It is suggested that light oils are examined in some type of kinematic viscometer and the results converted to Saybolt values.

ASTM DESIGNATION: D 88–56

Scope

1. This designation covers procedures for the empirical measurement of Saybolt viscosity of petroleum products at specified temperatures between 70° and 210°F. There is also included a special procedure for waxy and resinous substances.[1]

Definitions

2. (*a*) *Saybolt Universal Viscosity*—is the efflux time in seconds of 60 ml of sample flowing through a calibrated Universal orifice under specified conditions.

---

[1] A fundamental and preferred method for measuring viscosity is by use of kinematic viscometers as outlined in ASTM Method D 445, Test for Kinematic Viscosity. This method requires smaller samples, less time, and gives greater accuracy.

Saybolt Universal and Saybolt Furol viscosities may be obtained from kinematic viscosity values by the use of conversion tables given in ASTM Method D 446, conversion of kinematic viscosity to Saybolt Universal viscosity, and ASTM Method D 666, conversion of kinematic viscosity to Saybolt Furol viscosity, respectively.

Viscosity index calculation from kinematic viscosity is recommended over viscosity index calculations using Saybolt Universal values.

Saybolt Furol viscosity of asphaltic materials at high temperatures is covered by ASTM Method E 102.

(*b*) *Saybolt Furol Viscosity*—is the efflux time in seconds of 60 ml of sample flowing through a calibrated Furol orifice under specified conditions. The Furol viscosity is approximately one tenth of the Universal viscosity, and is recommended for those petroleum products which have viscosities greater than 1000 sec (Saybolt Universal), such as fuel oils, and other residual substances. The word "Furol" is a contraction of fuel and road oils.

Outline of Method

3. The efflux time in seconds of 60 ml of sample, flowing through a calibrated orifice, is measured under carefully controlled conditions. This time is corrected by an orifice factor and reported as the viscosity of the sample at that temperature.

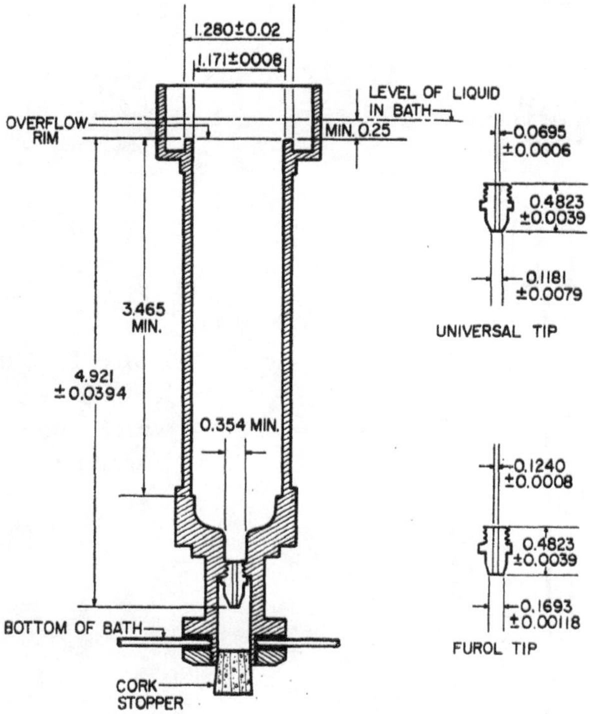

FIGURE 1. Saybolt Viscometer with Universal and Furol Orifice

Apparatus

4. (*a*) *Saybolt Viscometer and Bath,* as described in Appendix 1 and illustrated in Figure 1.

(*b*) *Withdrawal Tube,* as shown in Figure 2.

(*c*) *Thermometer Support,* as shown in Figure 3.

(*d*) *Filter Funnel,* as shown in Figure 4.

(*e*) *Receiving Flask,* as shown in Figure 5.

(*f*) *Timer,* graduated in tenths of a second, and accurate to within 0.1 per cent when tested over a 60-minute interval. Electric timers are acceptable if operated on a controlled-frequency circuit.

(*g*) *Viscosity Thermometers,* for reading the test temperature of the sample. The ASTM Saybolt Viscosity Thermometers shown in Table 1 shall be used.

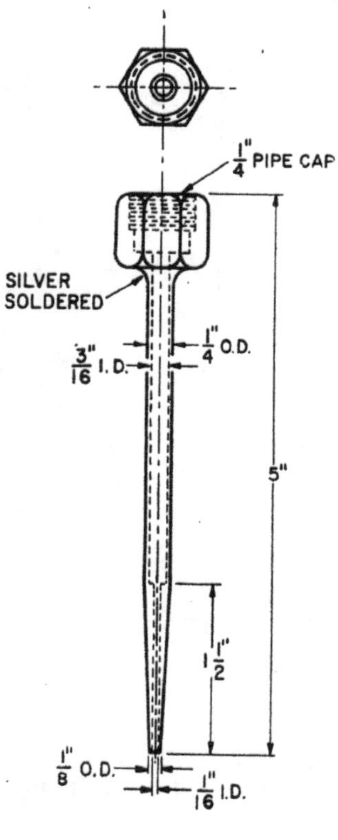

FIGURE 2. Withdrawal Tube for Use with Saybolt Viscometer

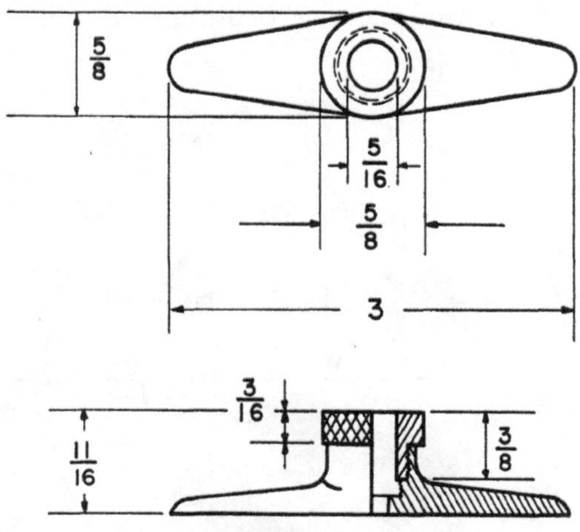

FIGURE 3. Thermometer Support

TABLE 1. ASTM Saybolt Viscosity Thermometers

Standard Test Temperature, °F	Temperature Range, °F	Subdivisions, °F	ASTM Thermometer No.
70	66 to 80	0.2	17 F
77	66 to 80	0.2	17 F
100	94 to 108	0.2	18 F
122	120 to 134	0.2	19 F
130	120 to 134	0.2	19 F
140	134 to 148	0.2	20 F
180	174 to 188	0.2	21 F
210	204 to 218	0.2	22 F

(*h*) *Bath Thermometers*—Viscosity thermometers, or any other means of equivalent accuracy.

Preparation of Apparatus

5. (*a*) Use a Universal orifice for lubricants and distillates with efflux times greater than 32 sec to give the desired accuracy.

Liquids with efflux times over 1000 sec are not conveniently tested with this orifice.

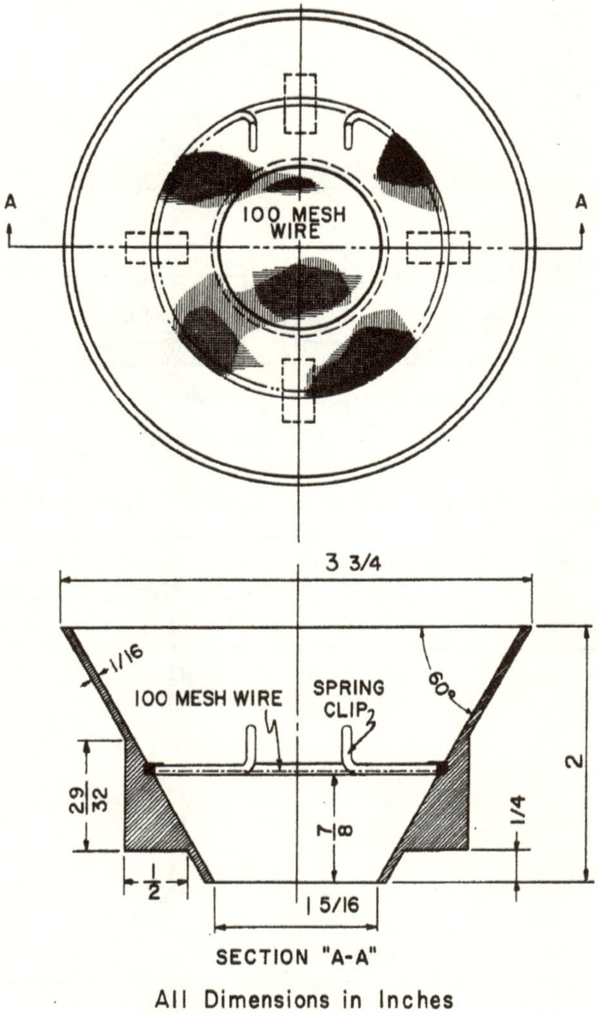

FIGURE 4. Filter Funnel for Use with Saybolt Viscometer

(*b*) Use a Furol orifice for residual matter with efflux times greater than 25 sec. The Furol efflux time is approximately one tenth of the Universal efflux time.[2]

[2] The Universal orifice is used at 70, 100, 130, and 210°F. The Furol orifice is used at 77, 100, 122, and 210°F.

(*c*) Set up the viscometer and bath where they will be free from drafts and rapid changes in air temperature. Locate them so that the sample will not be contaminated by dust or vapors during the test.

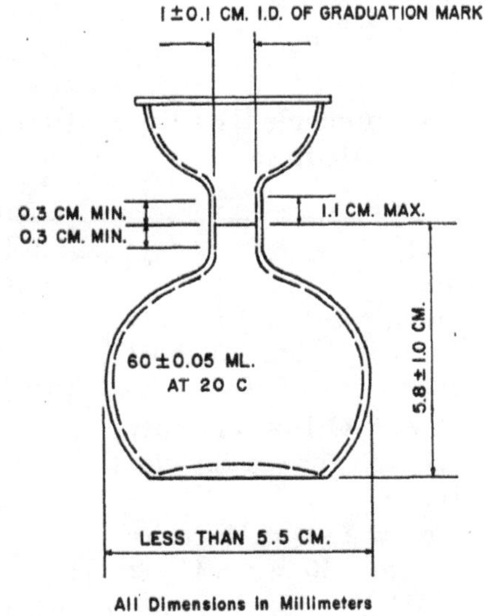

FIGURE 5. Receiving Flask.

(*d*) Viscosity determinations shall not be made at temperatures below the dew point of the room's atmosphere. Room temperatures up to 100°F will not introduce errors in excess of 1.0 per cent. For standardization and referee tests, the room temperature shall be kept between 68 and 86°F, and the actual temperature recorded.

(*e*) Fill the bath at least ¼ in. above the overflow rim of the viscometer. Table 2 lists recommended bath media for each test temperature.

(*f*) Provide adequate stirring and thermal control for the bath so that the sample will not fluctuate more than ±0.05°F after reaching the test temperature.

(*g*) Clean the viscometers with an effective nontoxic solvent and remove all solvent from the gallery and viscometer.[3]

[3] The plunger commonly supplied with the viscometer should never be used for cleaning, as the overflow rim and walls of the viscometer may be damaged by its use.

TABLE 2. Bath Test Temperatures

Standard Test Temperature, °F	Recommended Bath Medium	Maximum Temperature Differential,[a] °F	Functional Precision
70	Water	±0.1	±0.05
77	Water	±0.1	±0.05
100	Water or Oil of Viscosity 50 to 70 sec S.U.[b] at 100°F	+0.25	±0.05
122	Water or Oil of Viscosity 120 to 150 sec S.U. at 100°F	+0.35	±0.05
130	Water or Oil of Viscosity 120 to 150 sec S.U. at 100°F	+0.5	±0.05
140	Water or Oil of Viscosity 120 to 150 sec S.U. at 100°F	+1.0	±0.1
180	Water or Oil of Viscosity 330 to 370 sec S.U. at 100°F	+1.5	±0.1
210	Oil of Viscosity 330 to 370 sec S.U. at 100°F	+2.0	±0.1

[a] Maximum difference allowed between bath temperature and test temperature to maintain thermal equilibrium while stirring sample in viscometer with test thermometer.

[b] S.U. = Saybolt Universal.

Calibration of Viscometers [4]

6. (*a*) Calibrate the Saybolt Universal viscometer at periodic intervals by measuring the efflux time at 100°F of an appropriate

[4] Saybolt viscometers can be certified at the National Bureau of Standards by the use of oil standards having viscosity values established, using distilled water at 68°F as the basic standard.

viscosity standard, following the procedure given in Section 7. See Appendix 2.

(*b*) The efflux time shall equal the certified Saybolt viscosity value. If the efflux time differs from the certified value by more than 0.2 per cent, calculate a correction factor, *F*,[5] for the viscometer as follows:

$$F = \frac{V}{t}$$

where:

V = certified Saybolt viscosity of the standard, and
t = efflux time in seconds at 100°F.

(*c*) Calibrate the Saybolt Furol viscometer at 122°F in the same manner as the Saybolt Universal, using a viscosity standard having a minimum efflux time of 90 sec.

(*d*) Viscometers or orifices which have corrections in excess of 1.0 per cent shall not be used for referee testing.

Procedure

7. (*a*) If the test temperature is above room temperature, the test may be expedited by preheating the sample to not more than 3°F above the test temperature. Never preheat any sample to within 50°F of its flash point, as volatility losses may alter its composition.

(*b*) Insert a cork stopper, having a cord attached for its easy removal, into the air chamber at the bottom of the viscometer. The cork shall fit tightly enough to prevent the escape of air, as evidenced by the absence of oil on the cork when it is withdrawn.

(*c*) Filter the prepared sample through a 100-mesh screen directly into the viscometer until the level is above the overflow rim.

(*d*) Stir the sample until its temperature remains constant within 0.05°F of the test temperature during 1 minute of con-

[5] The correction factor applies at all viscosity levels and for all temperatures, provided the calibration is based on a viscosity standard having an efflux time between 200 and 600 sec.

tinuous stirring.[6] Stir with a viscosity thermometer equipped with a thermometer support (Figure 3). Use circular motion at 30 to 50 rpm in a horizontal plane.

(*e*) Remove the thermometer from the sample. Quickly remove the oil from the gallery until its level is below the overflow rim. This is done by placing the tip of the withdrawal tube (Figure 2) at one point in the gallery and applying suction. Do not touch the overflow rim with the withdrawal tube, or the effective head of the sample will be reduced.

(*f*) Place the receiving flask (Figure 5) where the stream of oil from the bottom of the viscometer will just strike the neck of the flask. The graduation mark on the flask shall be between 10 and 13 cm from the bottom of the viscometer tube.

(*g*) Snap the cork from the viscometer using the attached cord. At the same instant, start the timer. Stop the timer the instant the bottom of the meniscus reaches the graduation mark. Record the efflux time in seconds.

Special Procedure for Waxy or Resinous Substances

8. (*a*) Steam-refined cylinder oils, black lubricating oils, residual fuel oils, and other matter which may possibly contain waxes or resins require special handling to obtain reproducible results. In order to control the possible anomalies in testing such matter, the thermal history of the sample shall be controlled. This special procedure will aid in the control of wax crystals and lattice-structure formations.

(*b*) Heat the filtered sample in a loosely stoppered container for 1 hour in an oven maintained at 212° to 215°F.

(*c*) Immediately pour the sample into the clean, stoppered viscometer without refiltering. Place a thermometer with support into the sample. Do not disturb the sample until it is 1°F above test temperature.

(*d*) When the sample reaches this temperature, stir it with the thermometer, using circular motion at 30 to 50 rpm, until the temperature of the sample remains constant within 0.05°F of the test temperature during 1 minute of continuous stirring.

---

[6] Never adjust the temperature by immersing hot or cold bodies into the sample. Such thermal treatment may affect the sample and the precision of the test.

(*e*) Complete the test as described in Section 7 (*e*), (*f*), and (*g*).

(*f*) Viscosity determinations of these substances shall be completed within 1 hour after filling the viscometer with sample. Do not run repeat determinations on the same portion of these substances.

Calculation and Report

9. (*a*) Multiply the efflux time by the calibration factor for the viscometer used.

(*b*) Report the corrected time in seconds as the Saybolt Universal viscosity or Saybolt Furol viscosity of the oil at the temperature at which the test was made.

(*c*) Report the values below 200 sec to the nearest 0.1 sec. Report all values of 200 sec or higher to the nearest whole second.[7]

APPENDIX 1

Saybolt Viscometer and Accessories

A1. *Viscometer*—The viscometer, illustrated in Figure 1, shall be constructed entirely of corrosion-resistant metal, conforming to dimensional requirements shown in Figure 1. The orifice tip, Universal or Furol, may be constructed as a replaceable unit in the viscometer. Provide a nut at the lower end of the viscometer for fastening it in the bath. Mount vertically in the bath and test the alignment with a spirit level on the plane of the gallery rim. Provide a cork or other suitable means to prevent the flow of sample until the start of the test; a small chain or cord may be attached to the cork to facilitate rapid removal.

A2. *Bath*—The bath serves both as a support to hold the viscometer in a vertical position and as the container for the bath medium. Equip the bath with effective insulation and with an efficient stirring device. Provide the bath with a coil for heating and cooling and with thermostatically controlled heaters capable of maintaining the bath within the functional precision given in Table 2. The heaters and coil should be

[7] Results in different laboratories with different operators under referee or standardization conditions of testing should not differ by more than 0.5 per cent.

located at least 3 in. from the viscometer. Provide a means for maintaining the bath medium at least ¼ in. above the overflow rim. The bath media are given in Table 2.

APPENDIX 2

Viscosity Standards

A3. Viscosity standards are available from two sources. These standards may be used with equal confidence provided they are used immediately after opening and not stored for reuse as permanent viscosity standards.

A4. *API Viscosity Standards*—(*a*) American Petroleum Institute liquid viscosity standards having certified Saybolt viscosity values established by annual cooperative determinations of kinematic viscosity values. The kinematic viscosity values are converted to Saybolt Universal and Furol viscosity values by means of conversion tables given in ASTM Methods in D 446 and D 666, respectively.

(*b*) The "alpha" standard has a Saybolt Universal viscosity of approximately 300 sec at 100°F. The "beta" standard has a Saybolt Furol viscosity of approximately 90 sec at 122°F. Exact values are supplied with each sample.

TABLE 3. Saybolt Viscosity Standards (Approximate Values)

Designation	Universal, Seconds At 100° F	Universal, Seconds At 130°F	Universal, Seconds At 210°F	Furol, Seconds At 122°F
Alpha	305	...	53	...
Beta	1810	...	128	91
Gamma	37	...	...	...
SB	300	...	...	...
SC	...	300	...	...
SF	...	...	...	200

A5. *NBS Viscosity Standards*—National Bureau of Standards liquid viscosity standards having accurate values supplied with each sample. Standard SB has a Saybolt Universal viscosity of approximately 300 sec at 100°F. Standard SF has a Saybolt Furol viscosity of approximately 200 sec at 122°F.

A6. *Routine Calibrations*—The viscosity standards may also be used for routine calibrations at other temperatures as shown in Table 3.

A7. *Other Viscosity Standards*—Other reference liquids, suitable for routine calibrations, may be established by selecting stable oils covering the desired range and determining their viscosities in a viscometer calibrated with an API or NBS standard as described in Section 6.

DETERMINATION OF THE KINEMATIC VISCOSITY OF PETROLEUM PRODUCTS BY GLASS VISCOMETERS

Kinematic viscosity may be determined by several types of instruments developed by various investigators. Unfortunately, no single instrument available is entirely satisfactory for *all* types of products. The Enjay Laboratories have investigated this subject rather thoroughly and have adopted five different models for common use; by proper selection any type of petroleum product can be successfully tested. The various models are all illustrated in Figure 1 and a brief description of each will be found in the following.

This method may be applied to any product which is a true viscous liquid at the temperature of the test. It is not applicable to products which contain extraneous matter or are volatile at test temperature.

Apparatus

(1) *Viscometers*—The various types of viscometers are available from commercial laboratory-supply houses. They may also be obtained through the Enjay Co., Inc., referring to the numbers listed in Table 1.

In each case, the size number listed in the first column of Table 1 will be the "C" factor of the instrument in question. The nonstarred items are all *unity factor* viscometers, i.e., the actual "C" factor will be within ±0.1 per cent of the value listed in the first column. The starred items are somewhat more difficult to manufacture, and the "C" factor cannot be controlled within such narrow limits. Of the starred items, the three SIL-VIS viscometers will all have "C" factors within ±0.2 per cent of the listed values, while the remaining ones will be within 10 per cent.

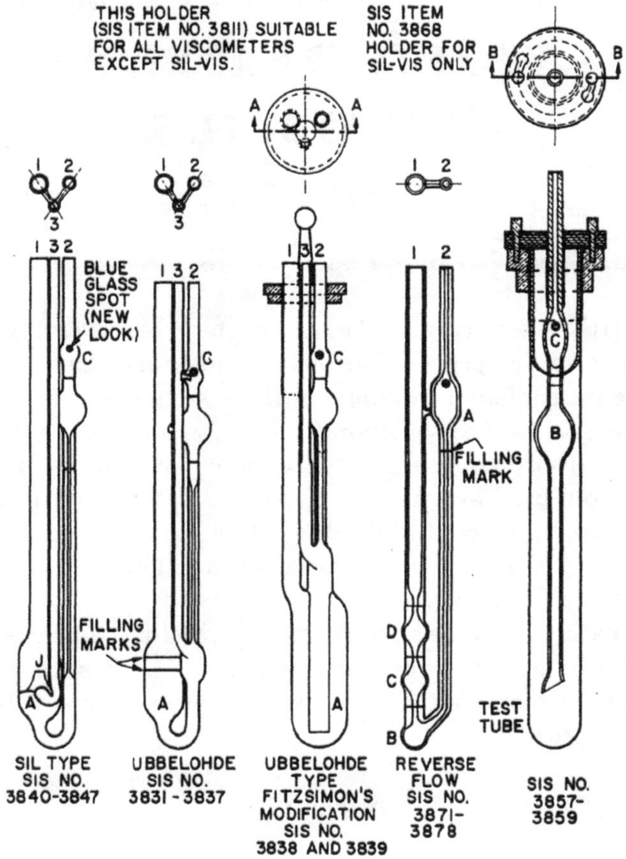

FIGURE 1. Types of Viscometers

Whatever the type, each viscometer will be accompanied by a certificate showing the exact numerical value of the "C" factor. For the nonstarred items, the acuracy of standardization is guaranteed to be within ±0.1 per cent of the certified values. In all other cases, the guaranteed accuracy is within ±0.2 per cent (3840 is accurate to 0.1 per cent).

These instruments are all calibrated on the basis of the new value for water as established by the United States Bureau of Standards and accepted by the ASTM as of July 1, 1953. To distinguish clearly between these viscometers and those calibrated on the basis of the value used prior to July 1, 1953, all instruments calibrated on the new basis have a blue dot fused into the glass near the top.

TABLE 1. Enjay Numbers of Viscometers

Size	Ubbelohde	Enjay (S.I.L.)	Enjay (SIL-Vis.)	Fitz-Simons	Reverse Flow
0.003	....	3840*	....	....	....
0.010	3831	3841	....	....	3871*
0.030	3832	3842	....	....	3872*
0.100	3833	3843	....	....	3873*
0.300	3834	3844	....	....	3874*
1.000	3835	3845	....	....	3875*
3.000	3836	3846	....	....	3876*
10.000	3837	3847	3857*	....	3877*
30.000	....	....	3858*	3838*	3878*
100.000	....	....	3859*	3839*	....

(2) *Holder*—Whatever the type in use, it is important that the viscometer is mounted in the bath in an absolutely vertical position. A satisfactory holder is shown in Figure 1 under SIS Item 3811. This holder will accommodate any type of instrument except the SIL-Vis viscometer, for which a special holder is available, as shown in Figure 1 under SIS Item 3868. This will mount the viscometer vertically in the test tube, and the tube in the bath.

Holder, complete, regular type	SIS ITEM 3811*
Handle only, for the above holder	SIS ITEM 3819*
Holder, special, for SIL-Vis pipet	SIS ITEM 3868*
Test tube, for SIL-Vis pipet	SIS ITEM 3865*

* SIS items are available from Enjay Co., Inc.

(3) *Cleaning Device*—The apparatus shown in Figure 2 will be very helpful in cleaning the viscometers. After a test is completed, the instrument is inverted and suspended on a hook until most of the oil has drained out. The tube is then placed on one of the pegs of the cleaner and the knurled knob on the base of the jet turned in a clockwise direction. By pressing the button on the valve on the front of the machine, a stream of solvent is

forced through the instrument until the tubes are clean. When the button is released, a stream of clean dry air, which should be at a gage pressure of about 10 psi, is directed through the instrument until it is dry.

The preferred solvent is a mixture containing 60 per cent carbon tetrachloride and 40 per cent naphtha of a boiling range of 100° to 300°F. By installing the machine in a place where there is no fire or health hazard, straight naphtha can be used.

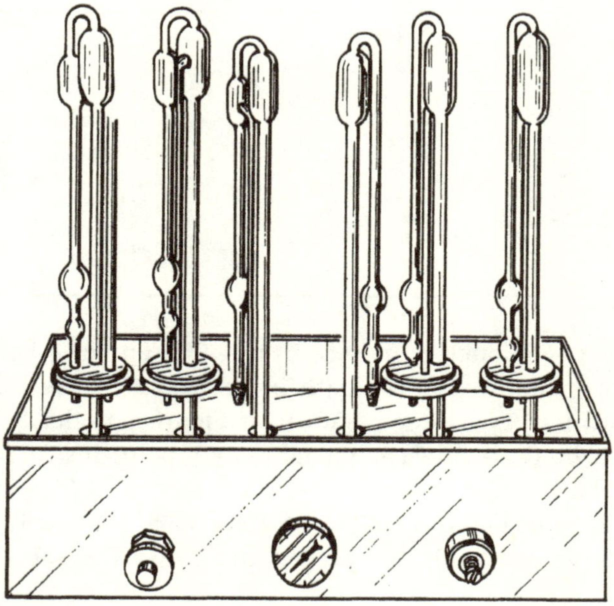

FIGURE 2. Viscometer Cleaning Apparatus

This device is not recommended for the cleaning of the SIL-VIS viscometers, which can be cleaned by immersion in a jar filled with a suitable solvent.

(4) *Bath*—For testing at temperatures of 70° to 210°F, the recommended bath may be obtained from Standard Inspection Supply (Enjay Co., Inc.). It consists essentially of a clear glass bowl, which is filled with water or some other transparent liquid. The bath is fitted with an efficient stirrer and can be held to within ±0.02°F of the desired temperature. The source of heat or refrigeration should not be less than 3 in. away from any viscometer when in place. When completely installed, the upper

bulb of the viscometer should be immersed to a depth of at least 1 cm below the surface of the bath liquid.

Kinematic viscosity apparatus, complete, including glass bowl and cover, cable and plug, thermo-regulator, and control box, but without viscometers and thermometers	
12 unit size, including 12 holders (Regular)	SIS ITEM 3800*
6 unit size, including 6 holders (Regular)	SIS ITEM 3802
Glass bowl only, without other equipment	
12 unit size	SIS ITEM 3801
6 unit size	SIS ITEM 3803

* SIS items are available from Enjay Co., Inc.

When testing at temperatures below that of the room, a special type will be required.

(5) *Timer*—Any convenient type may be used providing only that it is readable to at least 0.2 second, and accurate to ±0.1 per cent over a 60-minute period. Electric timers should be operated only on currents of carefully controlled frequencies.

(6) *Thermometers*—The available thermometers are listed in Table 2.

TABLE 2. Range, Selection and Number of Thermometers

Range °F	ASTM No. °F	For Tests at °F	SIS ITEM No.*
−2.5 to +2.5	72	0.0	84
−42.5 to −37.5	73	−40.0	85
−67.5 to −62.5	74	−65.0	86
−61.0 to −29.0	43	no specific temperature	89
+97.5 to 102.5	28	100.0	90
127.5 to 132.5	29	130.0	91
207.5 to 212.5	30	210.0	92
66.5 to 71.5	44	68.0 + 70.0	93
74.5 to 79.5	45	77.0	94
119.5 to 124.5	46	122.0	95
137.5 to 142.5	47	140.0	96
177.5 to 182.5	48	180.0	97

* SIS items are available from Enjay Co., Inc.

The thermometers should be checked for accuracy and any corrections appearing on the certificates applied in correcting the observed temperatures.

(7) *Strainers*—Before being tested, a portion of the oil should be filtered through a screen having two hundred openings to the linear inch. Two satisfactory strainers are available. One of these is flat, similar to the one commonly used for Saybolt viscosity; when this is used, the oil should be strained into a small beaker and then poured into the viscometer. The other is conical; when this is used, a small funnel is placed in the largest tube of the viscometer, the screen placed in the funnel, and the oil filtered directly into the instrument.

Strainer, 200 mesh, flat type, complete	SIS ITEM 3808
Gauze only	SIS ITEM 3815
Conical type, without funnel	SIS ITEM 3816

Preliminary Operation

(1) Before attempting to determine a viscosity, the operator must select the proper type and size of viscometer for the product under test. The following will be helpful in making this selection:

Type of Sample

(a) For opaque products, use the reverse-flow viscometer.

(b) For oils that are reasonably transparent, use either the SIL or the Ubbelohde types for ranges up to 10,000 centistokes, and the Fitz-Simons or SIL-VIS for more viscous oils. SIL-VIS should be used for test temperatures of 210°F or higher, while the Fitz-Simons is recommended for low-temperature work. The range of viscometers is listed in Table 3.

In any case, the size should be such that the minimum efflux time will be at least 100 seconds for routine testing and 200 seconds for precision work. Although not critical, it is suggested that the maximum efflux time is held to 1000 seconds.

(2) Whatever the type, it is important that the viscometer is chemically clean at the start of the test. At least once a week,

and preferably once a day, the instrument should be cleaned by rinsing with chromic acid, followed by rinsings with water and acetone, and dried by a current of clean dry air. In this operation the following should be observed:

TABLE 3. Range of Viscometers

Size of Instrument	Range in Centistokes
0.003	0.3 to 2
0.01	1 to 10
0.03 or 0.05	3 to 50
0.1	10 to 100
0.3 or 0.5	30 to 500
1.0	100 to 1000
3.0 or 5.0	300 to 5000
10.0	1000 to 10000
30.0	3000 to 30000
100.0	10000 to 100000

(a) IT IS IMPORTANT that the chromic acid is not allowed to remain in contact with the glass for more than about 10 minutes. Serious damage to the capillary may result if the contact time is much more than this.

(b) The oily matter must be removed from the inside of the instrument before the acid is added.

In most cases the oily matter may be removed by the use of the cleaning device listed under *Apparatus*. If carbon tetrachloride is to be used as the solvent, the apparatus *must* be mounted in a hood. The vapors of this solvent are toxic, and the exposure of the operator to these fumes should be kept at the minimum. If necessary, the viscometer may be given a preliminary cleaning by prolonged soaking in Varsol, followed by a rinse with a more volatile solvent. In any case, it is important that the viscometer is completely clean and dry before use. If the atmospheric conditions are such that the temperature of the viscometer falls below the prevailing dew point during the drying, the air should be heated.

(3) Heat or cool the bath to the proper temperature and immerse proper bath thermometer into it. The thermometer should

be so located that the top of the mercury column will be at least 1 cm below the surface of the liquid, but in no case should the expansion bulb at the top be submerged.

Each thermometer sold by the Standard Inspection Supply (Enjay Co., Inc.) will be accompanied by a certificate showing the correction if any to be applied at the test temperature. This correction should be noted, and the bath temperature adjusted accordingly. In addition, at least once each week, and preferably once a day, the bath thermometer should be checked by immersing the bulb and the lower portion of the stem in a slurry of clean cracked ice and noting the reading of the ice-point section. If this reading differs from the one given on the certificate for the thermometer in use, the listed correction for test temperature should be modified accordingly.

From here on, the procedure will depend on the type of instrument selected. Complete directions will, therefore, be given for a determination using the SIL type viscometer. Any special directions for handling the others will be listed in separate sections.

Procedure—SIL

(1) Fit the viscometer into a holder, put a clean funnel into tube 1 (see Figure 1), and put a screen into the funnel. Tilt the instrument slightly and filter a portion of the sample through the screen, collecting the oil in bulb A. Add enough sample so that the liquid level will be not less than $\frac{1}{4}$ nor more than $\frac{1}{2}$ in. above the opening J. In effecting this filling, it is important that bulb A is filled completely, without trapping any air in it.

(2) When bulb A is full, remove the funnel and put the viscometer into the bath previously brought to test temperature. Be sure that the viscometer is mounted in an absolutely vertical position. Allow it to remain in the bath until the sample has come to test temperature.

In the case of Diesel fuels or light lubricants, 10 to 15 minutes are usually sufficient. For heavier products, 20 to 30 minutes will be required.

(3) When the oil has come to temperature, apply gentle suction to tube 3 and remove any excess oil above the opening J. Allow the instrument to stand for 1 to 2 minutes, and repeat the removal of excess oil if necessary. After the excess has been

removed, apply gentle suction to tube 2 until the small bulb above the upper timing mark is about half full. Stop the suction and allow the oil to drop back toward bulb B.

When applying suction to tube 2, it is important that no positive pressure is placed on the tube, since this would remove a part of the test portion of the oil.

(4) When the level of the liquid reaches the upper timing mark, start a stopwatch or timer. When the level reaches the lower mark, stop the timer. Note and record the time in seconds.

If the time is less than 100 seconds, discard the test and repeat using a viscometer with a smaller capillary. For referee tests the minimum efflux time should be 200 seconds. If the time is above the permissive limit, repeat the test until two successive observations agree to within ±0.2 per cent. Average the two values, and use the mean for calculating the kinematic viscosity.

Procedure—Ubbelohde

(1) Fill the instrument by filtering the oil through a screen in a funnel which is mounted in the largest tube. Add enough sample so that the liquid level will be between the two marks on bulb A (see Figure 1).

(2) Mount the tube in the bath and allow to come to temperature. Put a finger over tube 3, apply suction to tube 2, fill the tube until bulb C is about half full, and determine the viscosity as previously described.

When using this instrument, it is important that after bulb B has been filled, the excess oil is permitted to break away from the lower end of the capillary before the timing is started. When the sample is a light oil, no special precautions are necessary. However, when the oil is rather heavy, the operator should put a finger over the top of tube 2 after the suction has been removed and allow it to remain there until the oil has broken away from the lower end of the capillary.

Procedure—Fitz-Simons

(1) Fill the lower bulb in the same way as for the Ubbelohde, adding enough sample so that the lower bulb will be about two thirds full. Put the viscometer into the bath and allow to come to temperature.

(2) When the sample is at test temperature, put a finger on tube 3, apply suction to tube 2, and fill bulb C about half full. Stop the suction and immediately place a finger above the opening of tube 2. Allow the finger to remain in place until the oil has dropped away from the lower end of the capillary. Complete the determination as previously described.

Procedure—SIL-VIS

(1) Collect the filtered sample in the test tube, adding enough oil to fill the tube to a depth of about 2 in. Put the proper size viscometer into the holder, and fit the holder around the mouth of the test tube, locating the viscometer so that its tip is at least 1 in. below the surface of the oil (see Figure 1). Put the test tube into the bath, and allow the sample to come to temperature.

When using this type, the time required for the sample to come to testing temperature will be much greater than with any of the others. When testing at 210°F, leave the tube in the bath for 30 to 40 minutes. At any lower temperature, this time will be much greater. The use of the SIL-VIS is, therefore, recommended only for testing at 210°F or higher.

(2) When the oil has come to temperature, apply gentle suction to the open end of the viscometer and fill bulb C about half full. Stop the suction and raise the viscometer until the lower tip is well above the surface of the remaining liquid in the test tube. Determine the viscosity in the usual manner.

The timing should not be started until most of the oil has drained away from the outer surface of the lower end of the viscometer. If necessary, stop the flow before the timing is started by placing a finger over the open end of the viscometer. Be sure that at the completion of the test the lower tip of the viscometer is still above the final level of the oil in the test tube.

The shape of the tip at the lower end has an important bearing on the rate of flow. If this tip becomes damaged or chipped while in service, the instrument should be discarded.

Procedure—Reverse Flow

(1) Collect the filtered oil at room temperature (70° to 90°F) in a small beaker. Connect the larger tube to the vacuum line,

invert the viscometer, and dip the open end of the smaller tube into the sample. When bulb A is filled to the filling mark, turn the viscometer to the horizontal position and carefully wipe off the excess oil without changing the volume of the test portion. Put the charged instrument into the holder and apply a short length of rubber tubing fitted with a pinch clamp to the open end of tube 2 (see Figure 1). Stop the flow of oil as the level reaches the lower end of the capillary. Mount the charged viscometer in the bath and allow to come to temperature.

When using this type of viscometer, the volume of the original sample is critical, and the instructions must be followed closely. In particular, the temperature of the sample must be 80 ± 10°F when the tube is filled originally. If the room temperature (and consequently that of the sample) is beyond the permissive limits, the sample must be warmed or cooled as necessary before the tube is filled.

(2) When the sample in the bath has come to test temperature, remove the rubber tube and allow the liquid to flow into bulbs B, C, and D. Start a timer when the level of the meniscus reaches the mark between bulbs B and C and stop it when the level reaches the mark between C and D. A check determination can be made on the one filling by starting a second watch when the first is stopped and stopping the second when the liquid reaches the mark above D.

The calculations are generally the same for this as for the other types of viscometers. However, it should be noted that the exact numerical value of the "C" factor will change with a change in test temperature, and also that the "C" for bulb C is not the same as for bulb D. The exact values of the "C" factor under all conditions will be found in the certificate accompanying the instrument, and should be used in all calculations.

Calculations

(1) The usual equation for calculating kinematic viscosity is:

$$K = Ct - B/t$$

where

C is the calibration constant ("C" factor).
B is the kinetic energy constant ("B" factor).

t is the running time in seconds.
K is the kinematic viscosity in centistokes.

Any viscometer sold by the Standard Inspection Supply will be accompanied by a certificate showing the exact numerical values of the B and C factors. The equation should be used for calculating the viscosity whenever the running time of the sample falls within the data in Table 4.

TABLE 4. Running Time of the Sample

SIL size	Running time less than seconds	Ubbelohde size	Running time less than seconds
0.003	600	0.01	400
0.01	275	0.03 or 0.05	230
0.03 or 0.05	150	0.10	125

Within the limits of Table 4, the influence of the "B" factor is appreciable, and must be taken into account in the calculations. Beyond the limits of the table, the influence of the "B" factor is negligible, and may be disregarded.

(2) For routine testing, when unity-factor viscometers are used, it is sufficiently accurate to multiply the observed running time by the size number of the viscometer whenever the test conditions are other than those in Table 4. However, the kinematic viscosity should be calculated by the equation $K = Ct$ whenever:

(a) a non-unity-factor viscometer is used (any reverse flow, SIS-Vis, or Fitz-Simons type, or the SIL size 0.003);
(b) a precision test is required.

Reporting

(1) When the specifications for the sample under test require a determination of kinematic viscosity, report the value found, the temperature of the test, and (to show that the determination was actually made in a kinematic viscometer) include the letter K.

A typical report might read "Kinematic viscosity, 29.6 centistokes at 100 K."

(2) In some cases, even though the specifications may require

a Saybolt viscosity, it is permissible to make the actual determination in the kinematic viscometer and to convert the results to Saybolt. In this case, convert the value to, and report as, Saybolt seconds, but include in parentheses the letter "K" to show that the determination was actually made with a kinematic viscometer.

A typical report might read "Saybolt viscosity, 120 seconds at 210 (K)."

CAUTION: *Be sure* to use tables of conversion based on the new kinematic viscosity of water.

The described procedure may also be used to obtain values in Engler and Redwood units. The final report should contain the letter "K," to show that the reported figures are converted values and not those actually obtained in the respective instruments.

L.I.C. No. 40.01C

COLOR OF PETROLEUM PRODUCTS BY THE TAG-ROBINSON COLORIMETER (COLOR STABILITY OF LUBRICANTS)

The Tag-Robinson colorimeter was originally developed by C. I. Robinson of the Standard Inspection Laboratory. The instrument is available from a number of different sources. The results obtained with this instrument are entirely empirical and dependent on the quality of the color glasses, which should be carefully selected and certified.

The Tag-Robinson colorimeter may be used for the examination of any petroleum product which has a color within the range of the instrument. Provision is made for determining color on the sample as received, or after a specified treatment.

Method

The instrument, shown in Figure 1, consists essentially of two large tubes with glass bottoms mounted on a support. Two smaller tubes are suspended from a common movable eyepiece. One side is used for the sample; the larger tube is filled with oil and the smaller (or oil-immersion) tube has a plain glass bottom. The other side is used as a reference; the large tube is empty and provision is made for attaching a standard color glass to the bottom of the smaller. When the operator looks through the eyepiece, the field of vision will be divided, with one side representing the sample and the other the standard. By raising or lowering the eyepiece, the operator will look through a greater or smaller depth of oil and the color of that side will be changed accordingly. When the colors of the two halves match, the reading of the gage on the back of the instrument is reported as the Tag-Robinson color.

The reading represents the height, in centimeters, of a column of oil through which the operator must look to obtain a color match. Although a reading of 9 (centimeters) is the maximum that can be noted on the scale, provision is made for extending this range by the use of standard glasses of different colors.

Apparatus

(1) *Colorimeter*—The complete instrument is shown in Figure 1.

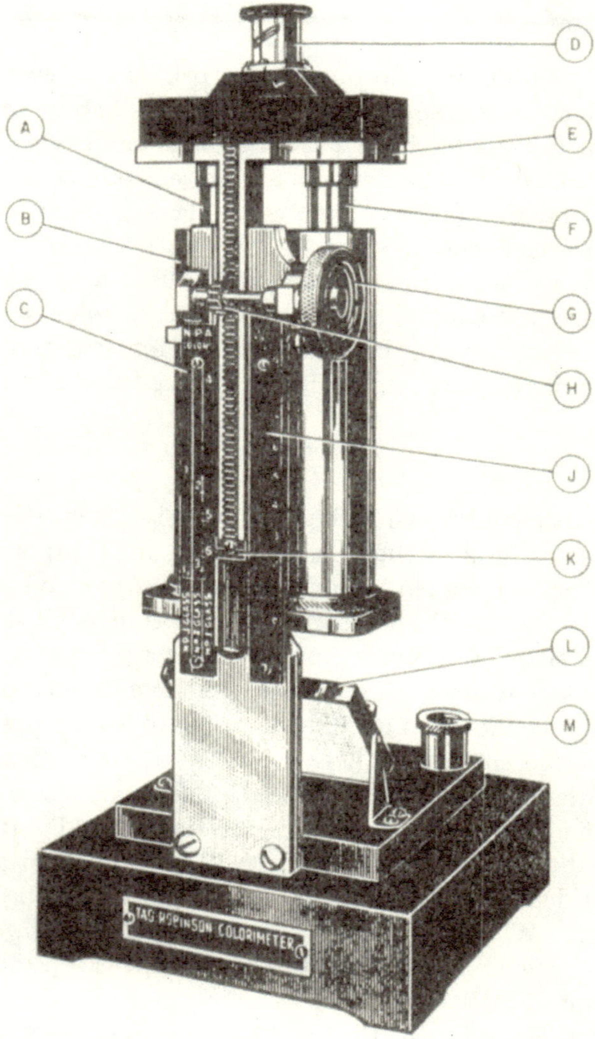

FIGURE 1. Tag-Robinson Colorimeter

(2) *Source of Light*—The light which enters the tube is normally reflected upward by a mirror (L in Figure 1). If daylight is used, it is important that there are no colored objects in the background. For referee or precision testing the light must be supplied by a daylight lamp, and interference by colored objects eliminated.

Procedure–Color of Liquids

(1) Remove the immersion tube and the sample tube (F and G in Figure 1) and clean each thoroughly with some light solvent naphtha. Wipe with a clean cloth or dry with air.

(2) Fill the sample tube partly with the sample under test and put it into position in the instrument. Insert the immersion tube and attach the upper part to the eyepiece platform.

Normally, the sample tube is filled to a point high enough to cover the lower cap of the immersion tube when this is raised to its highest position. If necessary, a smaller quantity may be used, but it is essential that the lower glass of the immersion tube is in the sample when a reading is taken.

(3) Remove the color-standard tube, attach color glass No. 1 at the bottom, insert the tube into the metal tube (B of Figure 1), and attach the upper portion to the platform.

(4) Look down through the eyepiece, and lower or raise the assembly as necessary until the colors of the two halves in the field of vision just match.

(5) If, but only if, the color cannot be matched with the No. 1 glass, remove it and replace with the No. 2 glass. If the color still cannot be matched, insert the No. 3 glass. If the color cannot be matched now either, the sample must be examined in some other instrument.

The given sequence is important. It is essential that to determine the color with the No. 1 glass, if possible, and to insert the No. 2 glass only when it is impossible to obtain a reading with No. 1. Similarly, No. 3 should be used only when the operator cannot obtain a match using No. 2. Occasionally, samples may be encountered whose color shade does not match the color of any glass disk. In this case the intensity of the light, rather than the exact tint of the oil, should be matched by turning the knob.

(6) When a match has been obtained, note and record the

reading of the scale on the back of the instrument and the number of the color glass used for the test. Determine the Tag-Robinson color by means of the Table 1.

TABLE 1

Color Glass	Color to Be Reported
No. 1	Scale reading
No. 2	8 plus the scale reading
No. 3	16 plus the scale reading

If the color of the sample is lighter than that of the No. 1 glass at a reading of 9, but darker at a reading of 1¼ with the No. 2 glass, consider the reading as 9+. Similarly, if light at 9 using the No. 2 glass but dark at 1¼ with the No. 3, report 17+.

(7) When specified, if the color of the sample is less than ¾, prepare a mixture containing 15 per cent by volume of the sample and 85 per cent of a kerosene having a Saybolt color not darker than 21+. Examine this mixture and report as "diluted Tag-Robinson color."

Reporting

(1) Report the reading of the scale, corrected as necessary for the color glass used, as the

"Tag-Robinson color .... units" or as "diluted Tag-Robinson color"

(2) If the oil as received is cloudy, include this information in the report. When specified, remove the cloud by filtering, examine the clear oil, and include this information also in the report.

Procedure–Color Stability

It is sometimes desirable to determine the color stability of certain products. In this case, determine the color of the sample as received in the normal manner. Then pour 100 ml of the sample into a clean, dry 4-ounce bottle, put the unstoppered bottle into an oven or bath maintained at 212° ± 2°F, and leave it undisturbed for 16 hours. Remove the bottle from the oven, allow to cool in a dark compartment, and determine the color again. Compare the results and report the difference as

color loss after 16 hours at 212°F . . . .

The bath may be either an air oven or a water bath. It is important that the limit of 212° ± 2°F is observed. If a liquid bath is used, the level of liquid should be above the neck of the sample bottle. If an oven is used, a dummy sample containing a thermometer should be put into the oven and the temperature of the thermometer noted at intervals.

Procedure—Waxes

The method is the same as for liquids, with the single exception that the sample must be heated to keep it in a molten condition. For this purpose a special tube to replace G in Figure 1 must be used. This tube has a built-in electrical heating unit and, when used with a voltage regulator, will keep the sample at any desired temperature.

To examine waxes, replace the usual tube with the special one, turn on the electric current, and allow the unit to warm up. In a separate container, heat the sample to some 30°F above its melting point, taking care to avoid local overheating, and pour the liquid into the special tube. Insert the regular immersion tube (F in Figure 1) and inspect the sample. If necessary, leave the current on until the sample is completely molten. Shut off the heat and, when convection currents in the liquid are no longer noted, conduct the test as usual.

The Tag-Robinson instrument normally has a small support, or divider, along the main shaft to hold the tubes firmly in place. The diameter of the electrically heated tube is somewhat greater than that of the normal one (G in Figure 1) and will not fit into place unless this divider is removed. However, the removal of the divider affects in no way the stability of the instrument or its operation.

The electrically heated tube is much more satisfactory than one with a water or steam jacket, since there is no possibility of leakage of water. Precautions must be taken, however, to avoid overheating the sample. The special tube should, therefore, be used with a voltage regulator, and the temperature set at not more than 30°F above the melting point of the sample.

At first glance, it would seem that a reading of 1 using the No. 2 glass would be the same as a reading of 9 with the No. 1. Actually this is not true. Colors may vary sharply in the overlap area. It is, therefore, essential for the operator to determine the color with the No. 1 glass if possible, and use No. 2 only for oils that are definitely lighter than a reading of 9.

COLOR OF WHITE PETROLEUM PRODUCTS BY THE SAYBOLT CHROMOMETER

The Saybolt chromometer was originally designed and intended to measure the color of new, unoxidized samples of kerosene and other similar white oils. The scope was gradually extended to cover paraffin waxes and products which had been artificially oxidized.

Complete instructions for the use of the Saybolt chromometer are given in ASTM D 156, which is reprinted here. Appropriate instructions for the various modifications will be outlined as necessary.

It is preferable to use a special stand for mounting a daylight lamp directly under the chromometer to exclude outside sources of light. When it is used, the frosted mirror should be removed from the bottom of the chromometer and the rest of the apparatus placed directly on top of the stand. Since the daylight lamp will shine directly into the color tubes, any interference from outside sources of light will be reduced to a minimum. While not mandatory, it is strongly recommended that this stand is included in the assembly, particularly for referee testing.

The test is entirely empirical and the final results are dependent, to a large degree, on the color glasses used in the instrument. It is, therefore, recommended that all items are properly certified.

The modifications to the usual procedure are as follows:

Kerosene, White Oil, Etc.

Follow ASTM D 156 without exception.

Paraffin Wax

Some provision must be made for keeping the wax in a molten condition, either by an electrical heater as described in D 156, or by the Pyrex tube shown in Figure 2. When the tube is used, hot water is circulated through the jacket, the temperature of the water being not over 10°F above the melting point of the wax.

Color Stability of Kerosene

Fill two 4-oz sample bottles with the sample under test, insert the corks loosely, and put the bottles into a constant-temperature bath previously brought to the test temperature of 212°F. After ½ hour, secure the corks in place and leave the bottles undisturbed in the bath for a total of 16 hours at the test temperature of 212°F. At the end of 16 hours, remove the bottles, cool, and determine the color of the oxidized oil as usual.

If the colors of the two portions differ by more than four units, repeat the entire procedure. If the colors do not differ by more than four units, average the two results and report the mean as

Saybolt color stability .... units

ASTM DESIGNATION: D 156–53 T

Scope

1. This method describes a procedure for determining the color of refined oils, such as undyed motor and aviation gasoline, jet-propulsion fuels, naphthas and kerosene, and, in addition, petroleum waxes and pharmaceutical white oils.

Definition

2. *Saybolt Color* is a number related to the depth of a column of material, the color of which is compared with specified glass standards. The range of numbers is +30 to −16; the greater the number, the lighter the color.

Outline of Method

3. The height of a column of sample is decreased by levels corresponding to color numbers until the color of the sample is

unmistakably lighter than that of the standard. The color number above this level is reported, regardless of whether the sample was darker, questionable, or a match at the higher level.

Apparatus

4. The Saybolt chromometer consisting of sample and standard tubes, optical system, light source, and color standards.

Standardization of Apparatus

5. (*a*) Remove the glass disk from the bottom of the oil tube. Clean the disk, oil tube, and plain tube. When deposits are not removable by wiping or solvent rinsing, wash with soap and water. After cleaning, rinse with distilled water and with acetone or some other suitable solvent, and dry. Assemble the oil tube, and position the tubes in the instrument.

(*b*) Using the specified light source and illumination, observe the comparative light intensity of the two halves of the optical field, with both tubes empty, and with the 12-mm diaphragm removed from under the plain tube. The intensity of light observed in each half of the optical field must be the same. Adjustment in the position of the light source may be necessary to achieve this match.

(*c*) Replace the 12-mm diaphragm under the plain tube, and fill the oil tube to the 20-in. mark with distilled water. The intensity of the light observed in each half of the optical field must be the same, for the instrument to be judged satisfactory for use.

Preparation of the Sample

6. (*a*) If the sample is turbid, filter through a sufficient number of qualitative filter papers (Whatman No. 4, or equivalent) until it is clear.

(*b*) When preparing paraffin wax for testing, do not heat excessively, because oxidation may occur, with consequent discoloration of the sample.

Procedure for Refined Light Oils and Pharmaceutical White Oils

7. (*a*) Flush the oil tube with a portion of the sample, allowing the tube to drain thoroughly. Fill the oil tube with the sample [1] and compare with a whole color standard. If the sample is lighter than the color standard, remove the standard and replace it with a half standard. If the sample is darker than the single whole standard at 6¼ in., add another whole standard.

(*b*) With the proper color standard or standards in place, and the sample in the oil tube at a level where its color is decidedly darker than that of the color standard, draw off the sample slowly by means of the pet cock until the oil appears just slightly darker than the color standard. From this point, draw the sample level down to the nearest depth corresponding to color number as shown in Table 1. If the color of the oil observed through the eyepiece is still darker than the color standard, draw the oil down to the next depth given in Table 1, and compare. Continue this operation until a depth is reached where the sample and color standard match, or show questionable differences.[2] At this point, lower the oil column to the next specified depth and, if the oil is unmistakably lighter than the color standard, record the color corresponding to the next higher level as the Saybolt color.

Procedure for Petroleum Wax

8. (*a*) Heat the wax sample 15° to 30°F above its melting point. Preheat the oil tube.

(*b*) Pour the liquid wax into the oil tube; turn the heating element off, and, after the heat waves in the sample can no longer be noted, obtain the required readings as directed in Section 7.

---

[1] It is important that all samples in the color tube are free from air bubbles.

[2] Experience in the use of this instrument will obviate the necessity of following the step-by-step procedure outlined under (*b*) for choosing the proper color standards for each sample. Examples of the procedure are given in Table 2.

TABLE 1. Saybolt Colors Corresponding to Depths of Oil

Number of Color Standards	Depth of Oil, Inches	Color Number	Number of Color Standards	Depth of Oil, Inches	Color Number
One-half	20.00	+30	Two	6.00	+6
One-half	18.00	+29	Two	5.75	+5
One-half	16.00	+28	Two	5.50	+4
One-half	14.00	+27	Two	5.25	+3
One-half	12.00	+26	Two	5.00	+2
One	20.00	+25	Two	4.75	+1
One	18.00	+24	Two	4.50	0
One	16.00	+23	Two	4.25	−1
One	14.00	+22	Two	4.00	−2
One	12.00	+21	Two	3.75	−3
One	10.75	+20	Two	3.625	−4
One	9.50	+19	Two	3.50	−5
One	8.25	+18	Two	3.375	−6
One	7.25	+17	Two	3.25	−7
One	6.25	+16	Two	3.125	−8
Two	10.50	+15	Two	3.00	−9
Two	9.75	+14	Two	2.875	−10
Two	9.00	+13	Two	2.75	−11
Two	8.25	+12	Two	2.625	−12
Two	7.75	+11	Two	2.50	−13
Two	7.25	+10	Two	2.375	−14
Two	6.75	+9	Two	2.25	−15
Two	6.50	+8	Two	2.125	−16
Two	6.25	+7			

TABLE 2. Example of Procedure

Observation	Using One Whole Color Standard	Using Two Whole Color Standards
Oil darker at depth of	16 in.	4.5 in.
Oil darker at depth of	14 in.	4.25 in.
Oil questionable at depth of	12 in.	4.0 in.
Oil lighter at depth of	10.75 in.	3.75 in.
Saybolt color	+21	−2

Report

9. Report the recorded color units as "Saybolt color ......"; if the sample has been filtered, add the words "(sample filtered)."

Precision

10. The following criteria should be used for judging the acceptability of results:

(*a*) Duplicate results by the same operator should be considered suspect if they differ by more than one color unit.

(*b*) The results submitted by one laboratory should be considered suspect if it differs from that of another laboratory by more than two color units.

DETERMINING THE COLOR OF SOLVENTS BY THE PLATINUM-COBALT SYSTEM

This method is frequently used in measuring the color of products which are reasonably close to water white. It offers the distinct advantage that no special instrument is required; any well-equipped laboratory can make the necessary solutions and obtain reproducible results.

This method is based on ASTM D 1209, and was formerly part of D 268.

This method is not generally used on petroleum products, but is frequently specified in the examination of lacquer solvents, diluents and high-purity petrochemicals.

Apparatus

(1) *Nessler Tubes*—A quantity of 100-ml Nessler tubes will be needed. They should have a colorless bottom, and the sides should be covered with opaque paper to eliminate side light. Provision should be made for mounting these tubes in a vertical position in some sort of holder having a white background beneath the bottom of the tube.

Reagents

(1) *Distilled Water*—It is vitally important that the water used for the solutions is free of colored matter. If there is any doubt on this subject, the water should be redistilled.

(2) *Standard-Color Solution*—Sufficient potassium platinic chloride ($PtCl_4 \cdot 2KCl$) to contain 0.5 g of platinum, and sufficient cobalt chloride ($CoCl_2 \cdot 6H_2O$) to contain 0.25 g of cobalt should be dissolved in 100 ml of concentrated hydrochloric acid and diluted to 1 liter with distilled water. This solution has a standard color of 500.

The exact composition of the two solid reagents, particularly the cobalt chloride, is always subject to question. Each reagent should, therefore, be analyzed, and salt sufficient to contain the required amount of each metal used for the solution.

(3) *Set of Standards*—The colors are expressed numerically. To prepare any particular solution, divide the numerical value of the required standard by 5; the quotient represents the number of milliliters of the standard color solution which, when diluted to 100 ml with distilled water, will yield the required standard.

Example: A standard color of 100 is wanted:
100 divided by 5 = 20. Dilute 20 ml of the standard color solution to 100 ml with distilled water.

Procedure

(1) Fill a clean, dry Nessler tube to the 100-ml mark with the sample under test. Support the tube in an upright position, with a white background under the bottom.

(2) If the specifications for the sample under test require that the sample should not be darker than a specified numerical value, another tube containing the proper standard solution should be placed near the one containing the sample, and the two colors compared. If the color of the sample is not darker than the specified value, the sample should be reported as "Passes."

(3) If the specifications require a determination of the color, with no specified maximum, various standard solutions should be prepared, and each compared with the sample under test. The numerical value of the closest matching standard should be reported as

"Platinum-cobalt color . . . ."

Other Methods

The specifications may read simply that the sample should be "water white." In this case, fill the reference tube with a solution containing 0.0030 g of chemically pure potassium dichromate in 1 liter of distilled water. If the color of the sample under test is not darker than that of the dichromate solution, report the sample as *water white.*

TABLE 1. Color Relation

Pt-Co	Saybolt	Pt-Co	Saybolt
0	+30.0	50	+19.2
5	28.8	60	17.2
10	27.9	70	15.2
15	26.7	100	10.0
20	25.6	150	3.0
25	24.4	200	−3.0
30	23.5	250	−8.0
35	22.3	300	−12.0
40	21.2	350	−16.0

A Saybolt color of +21 is usually considered *water white.*

This system was developed by Allen Hazen in 1892. In his original scale, the darkest of the standard color solutions was considered to have a numerical value of 5, and lighter solutions proportional parts of the darker. Somewhat later the system was investigated by the American Public Health Association and, with one minor change, adopted as a satisfactory method. The one change was that the darkest color was considered to have a numerical value of 500, rather than 5, and lighter colors proportional parts of 500. (A Hazen color of 1 would be the same as an APHA color of 100.)

Over the years it has become customary to use the APHA scale, rather than Hazen's original one. However, in some quarters the term "Hazen color" is still used, even though results are reported in the APHA classification. In American industry, the terms "Hazen color" and "APHA color" are synonymous. To avoid confusion on this subject, it is recommended that the procedure outlined here is referred to as the *platinum-cobalt color system.*

GARDNER'S METHOD OF DETERMINING THE COLOR OF SOLVENTS

The Gardner method is frequently specified for measuring the color of solvents, particularly those used in the paint and varnish industry. The method is part of D 154–50.

Apparatus

(1) *Color Comparison Tube*—The preferred tube is that used for the Gardner-Holdt viscosity, which is available from the Central Scientific Co., 1700 Irving Park Road, Chicago, Illinois, under their catalog number 27143. If not available, a 50-ml Nessler tube may be substituted.

Color Standards

Any one of the following may be used:

(1) Dissolve pure powdered potassium dichromate in sulfuric acid in the amounts shown in Table 1. Apply gentle heat, if necessary, to effect solution of the chromate. These solutions will not remain color-stable for more than a few days, and must be prepared fresh as needed.

(2) Prepare three standard solutions and mix them in the proportions listed in Table 1. The standard solutions are:

(*a*) *Ferric Chloride* containing 5 parts by weight of $FeCl_3 \cdot 6H_2O$ and 1.2 parts of HCl (1:17). Adjust to exact color equivalence of a freshly prepared solution containing 3 g potassium dichromate in 100 ml concentrated sulfuric acid (sp. gr. 1.84)

(*b*) *Cobalt Chloride* containing 1 part by weight of $CoCl_2 \cdot 6H_2O$ to 3 parts of HCl (1:17)

(*c*) *Hydrochloric Acid* (1:17)

(3) Secondary refernce standards may be obtained in the form of eighteen colored glass disks, which are set into a pair of larger,

TABLE 1

Gardner Color Standard No.	$FeCl_3$ Solution milliliters	$CoCl_2$ Solution milliliters	HCl Solution milliliters	$K_2Cr_2O_7$ grams/100 ml H_2SO_4
1	0.13	0.19	99.68	0.0039
2	0.19	0.29	99.52	0.0048
3	0.29	0.43	99.28	0.0071
4	0.43	0.65	98.92	0.0112
5	0.65	0.97	98.38	0.0205
6	1.00	1.30	97.70	0.0322
7	1.70	1.70	96.60	0.0384
8	2.50	2.00	95.50	0.0515
9	3.30	2.60	94.10	0.0780
10	5.10	3.60	91.30	0.1640
11	7.50	5.30	87.20	0.2500
12	10.80	7.60	81.60	0.3800
13	16.60	10.00	73.40	0.5720
14	22.20	13.30	64.50	0.7630
15	29.40	17.60	53.00	1.0410
16	37.80	22.80	39.40	1.2800
17	51.30	25.60	23.10	2.2200
18	100.00	—	—	3.0000

plastic disks which are mounted to rotate in a housing for holding the sample tube and glass disk in close and fixed proximity. The pair of disks and a comparator housing may be obtained as Set 6C5-VC from Hellige, Inc., 3718 Northern Boulevard, Long Island City, New York.

Procedure

Determine the color by comparison with the reference standards, conveniently by comparing the sample and standard in Gardner-Holdt viscosity tubes. Make the comparison at 25° ± 5°C by placing the tubes close together and looking through them against a background which is substantially equal in color to the northern sky.

Reporting

Report the color in terms of the nearest match in the set of eighteen standards. For example, if the sample matches the No. 4 standard, report as

Gardner color 4

L.I.C. No. 40.70

DETERMINATION OF THE ACID WASH COLOR OF INDUSTRIAL AROMATIC HYDROCARBONS

This method appears in the ASTM Standard Methods under the designation D-848. It is suitable primarily for benzene, toluene, xylene, and similar industrial aromatic hydrocarbons. A special procedure is included for the examination of heavy solvent naphtha.

The acid wash color is the color developed in the separated acid when a sample is agitated with sulfuric acid under prescribed conditions.

Apparatus

(1) *Bottles for Color Standards*—These should be clear and unblemished, clean, and of the French square, flat-bottomed type. They are made of flint glass, fitted with a glass stopper, and have a capacity of 31 to 33 ml when filled to the neck. The bottles should be numbered consecutively from 0 to 14.

Satisfactory bottles may be obtained from the Eimer and Amend Company, 633 Greenwich Street, New York City under their catalog number 7–020.

(2) *Test Bottles*—The test bottles are the same as those used for the color standards, except that each is marked by etching to show when the bottle contains 7 ml or 28 ml. Colored crayons or similar agents must not be used for marking the bottles.

The bottles used for the sample and the standards should be as nearly uniform as possible and should all be taken from the same batch. A marked difference in thickness of the glass walls may result in an observed difference of one to two shades of color.

Reagents

(1) *Sulfuric Acid,* reagent grade, 96 ± 0.5 per cent by weight.
(2) *Sulfuric Acid,* reagent grade, 78 ± 0.5 per cent by weight.

(3) *Dilute Hydrochloric Acid*—25 ml of the concentrated chemically pure acid should be mixed with 975 ml of distilled water. The resulting solution is used in the preparation of the stock solutions.

Cleaning of Glassware

(1) All of the glassware used for this test, particularly the bottles, should be cleaned with a warm chromic acid cleaning solution before use. The glassware is then rinsed with tap and distilled water until clean and dried in an oven for 1 hour at 105°C.

Preparation of Stock Solutions

(1) *Solution A*—59.50 g of nickle-free $CoCl_2 \cdot 6H_2O$ are dissolved in a portion of the dilute HCl listed under REAGENTS. When solution is complete, the volume is adjusted to exactly 1 liter, using more of the dilute HCl.

(2) *Solution B*—In a similar fashion, 45.054 g of $FeCl_3 \cdot 6H_2O$ are dissolved in, and made up to 1 liter with, the dilute HCl.

Both reagents, particularly the ferric chloride, are hygroscopic, so that the final strength of the solutions is questionable, no matter how carefully prepared. As a safety precaution it is suggested that the exact strength of each solution is determined quantitatively, and adjusted if necessary.

(3) *Solution C*—3.5 volumes of solution A are mixed with 36.5 volumes of solution B and 90 volumes of distilled water.

(4) *Solution D*—3.5 volumes of solution A are mixed with 36.5 volumes of solution B.

(5) *Solution E*—This is a saturated solution of potassium chromate at 21°C.

(6) *Solution F*—A saturated solution of potassium dichromate at 21°C is prepared, and a portion diluted with an equal volume of distilled water.

Preparation of Reference Color Standards

The various reference color-standard solutions are prepared from the stock solutions. In each case, the bottle should be rinsed three times with the proper solution, and filled so that ¼ in. of vapor space will be left below the neck. After filling, each bottle should be indelibly labeled and tightly sealed.

Standard No. 0—Distilled water.

Standard No. 1—Equal volumes of distilled water and solution C.

Standard No. 2—5.5 volumes of solution C mixed with 2 volumes of water.

Standard No. 3—Solution C.

Standard No. 4—Equal volumes of distilled water and solution D.

Standard No. 5—5.5 volumes of solution D mixed with 2 volumes of water.

Standard No. 6—Solution D.

Standard No. 7—5 volumes of solution E mixed with 2 volumes of water.

Standard No. 8—Solution E.

Standard No. 9—7 volumes of solution E mixed with ½ volume of solution F.

Standard No. 10—6.5 volumes of solution E and 1 volume of solution F.

Standard No. 11—5.5 volumes of solution E and 2 volumes of solution F.

Standard No. 12—Equal volumes of solutions E and F.

Standard No. 13—2 volumes of solution E mixed with 5 volumes of solution F.

Standard No. 14—Solution F.

Procedure at Room Temperature below 85°F

(1) A clean test bottle is rinsed twice with sulfuric acid (96 per cent), drained, and filled with acid to the 7-ml mark. Sufficient sample to bring the total volume up to the 28-ml mark is then added, and the stopper inserted.

(2) Holding a finger over the stopper, the bottle is shaken for 15 to 20 seconds. When shaking the bottle, each stroke should be of 5 to 10 in., and the rate should be 5 cycles per second. At the end of the shaking period, the bottle is allowed to stand away from sunlight in surroundings below 85°F for the required time.

(3) The time of standing depends on the type of product under

test. Samples which are in the first group should be allowed to stand exactly 15 minutes, while those in the second group should stand exactly 5 minutes.

The first group includes nitration-grade, industrial-grade, and industrial 90-degree benzene; nitration-grade and 5- or 18-degree xylene; and other more highly refined products.

The second group includes industrial-grade xylene and refined solvent naphtha.

(4) At the end of the standing period, the bottle should be gently inverted once or twice to obtain a uniform color in the acid layer. The bottle is then held against a white background, or against daylight, so that the color may be clearly noted. (If the color is to be examined in daylight, it is important that there are no colored objects in the background.) The oily layer above the acid should be examined, and any change over the original color noted. In making this observation, a haze or cloudiness should not be interpreted as a change in color.

The change in color of the oil layer is of importance only if the sample is in the first group; if the sample is in the second group, the color of the treated oily layer may be disregarded.

(5) The acid layer in the bottle is then compared with the various color standards until two of the standards are found, one of which is slightly lighter, and the other slightly darker, than the sample. The closer matching of these two should be reported, followed by a plus or a minus sign to indicate the approximation. As an example, if the color of the acid layer is slightly lighter than the No. 4 standard, the sample should be reported as 4−. In addition, if the hue of the acid layer is different from that of the standard, the color number should be followed by (X). Thus a report reading "No. 4− (X)" would mean that the acid color is slightly lighter than the No. 4 standard, and the hue is not the same.

If preferred, the comparison of the acid layer and the standards may be done by means of a suitable color comparator.

Procedure at Room Temperatures above 85°F

The procedure is exactly the same as previously described, with the following exceptions:

At the start of the test the sample, acid, referee color standards, and the test bottle should be brought to a temperature of 77° to 85°F.

During the shaking period, the sample bottle should be insulated in some convenient way or wrapped with a cloth.

During the settling period, the sample bottle should be immersed in a bath maintained at 77° to 85°F.

Special Procedure for Hi-Flash Solvent or Heavy Solvent Naphtha

The procedure is as previously described, with the following exceptions:

The strength of the sulfuric acid should be 78 ± 0.5 per cent, instead of 96 per cent.

The settling period after shaking should be 5 minutes.

The color of treated oil should be disregarded.

Samples shall be reported as passing the test only when the oil layer shows no change in color (where applicable) and when the acid layer is not darker than the reference standard color prescribed in the specifications.

A cloudiness or haze in the oil layer should not be interpreted as a change in color.

The difference in hue should be disregarded; it is only necessary to determine whether the color of the acid layer is lighter or darker than the closest matching standard.

For tests in which the hue of the acid layer is about the same as that of the standard, duplicate determinations should not differ by more than the fractional part of the number as expressed by the plus or minus sign. In order to obtain this precision, it is important that the directions are followed closely. In particular, during the final comparison, the sample and the standards must be at the same temperatures.

SULFUR BY LAMP (CLOSED SYSTEM)

The recommended method for determining sulfur in petroleum distillates not heavier than kerosene is the same as that described in the current version of ASTM D 90, subject to the qualifications outlined in the next paragraphs.

(1) The general plan of precipitating the sulfur as barium sulfate must be followed for referee testing. It must be followed whenever the sample contains compounds of chlorine, bromine, and/or nitrogen, since these elements will form acidic compounds which would interfere with a volumetric titration. However, when the expected weight of barium sulfate is of the order of a few milligrams, it is recommended that the actual amount is determined by means of a turbidimeter.

(2) If the sample does not contain these interfering elements, the amount of sulfur may be determined either by titration with standard caustic solution or by a conductometric procedure. In this case, the combustion medium should be a synthetic mixture of oxygen and carbon dioxide, rather than purified air.

(3) The blending procedure described in section 7 should be used only if other items of apparatus are not available. Aromatic solvents can normally be burned without blending by using a modified type of burner, a synthetic combustion medium consisting of oxygen and carbon dioxide, or both.

(4) Under certain circumstances, the method described may not give the true sulfur content, since it is subject to errors from the following sources:

(a) Elemental sulfur, if present, will not be burned completely. Errors from this source can be minimized by allowing the sample to burn almost to dryness before removing the burner from the chimney. *Just before* the flame goes out, the burner assembly should be removed from the chimney, the flow of air stopped, and the burner weighed without delay.

(b) Certain types of sulfur gases in the atmosphere may not be completely removed by the caustic wash. Errors from this source can be eliminated by using the synthetic combustion medium described in ASTM D 1266.

Discussion

Certain laboratories are still using the open-lamp sulfur method. Their reasons for doing this are:

(a) The apparatus and technique are much simpler.

(b) The sulfur content of local samples is normally well below the specification maximum and small errors can be tolerated.

This is permissible if the local laboratory is willing to assume full responsibility for the final result.

Reporting

To show which method was used on a given sample, include in the report some key symbol. Typical symbols are as follows:

Lamp Sulfur
- Gravimetric—L
- Nephelometric—LN
- Titration—LT
- Conductometric—LC
- Open System—LO
- Dilution Method—LD

ASTM DESIGNATION: D 90–55 T

Scope

1. (*a*) This method describes procedures for the determination of total sulfur in concentrations above 0.002 per cent by weight in liquid petroleum products.

(*b*) The direct-burning procedure (Section 6) is applicable to gasoline, kerosene, naphtha, and other liquids that can be burned completely and without smoking in a wick lamp. The blending procedure (Section 7) is applicable to aromatics, gas oils and distillate fuel oils, naphthenic acids, alkyl phenols, and many other materials that cannot be burned satisfactorily by the direct-burning procedure.

(*c*) Halogens, nitrogen, phosphorus, or other acid-forming elements do not interfere in the analysis of commercial gasolines. Elemental sulfur is not determined quantitatively by this procedure.

Outline of Method

2. The sample is burned in a closed system, using a wick-type lamp and an atmosphere of sulfur-free air. The oxides of sulfur produced are absorbed and oxidized to sulfuric acid by means of hydrogen peroxide and determined gravimetrically as barium sulfate.

Apparatus

3. (*a*) *Chimneys, Absorbers, Spray Traps, and Lamps* (Figure 1), as specified in the Appendix.

(*b*) *Cotton Wicking*—Clean, unused, uniform, two-strand twisted cotton wicking of good quality, weighing 5 to 6 mg per cm per strand.

(*c*) *Vacuum System*—A vacuum manifold equipped with a vacuum regulating device, valves, etc., is shown in Figure 2, but any other similar system may also be used. The system shall be connected to a vacuum pump of sufficient capacity to permit a steady gas flow of 3 liters per minute through each absorber and to maintain a constant manifold pressure of approximately 40 cm of water below atmospheric.

(*d*) *Air-Purifying System*—A device to supply purified air to the burner manifold at a nearly constant pressure of approximately 20 cm of water and to the chimney manifold at a pressure of 1 to 2 cm of water. A convenient arrangement is shown in Figure 3, but any other similar system may also be used. The tubing which connects the chimneys to the manifold shall have an internal diameter not smaller than 0.25 in. (0.63 cm) in order to prevent unnecessary restriction of gas flow.

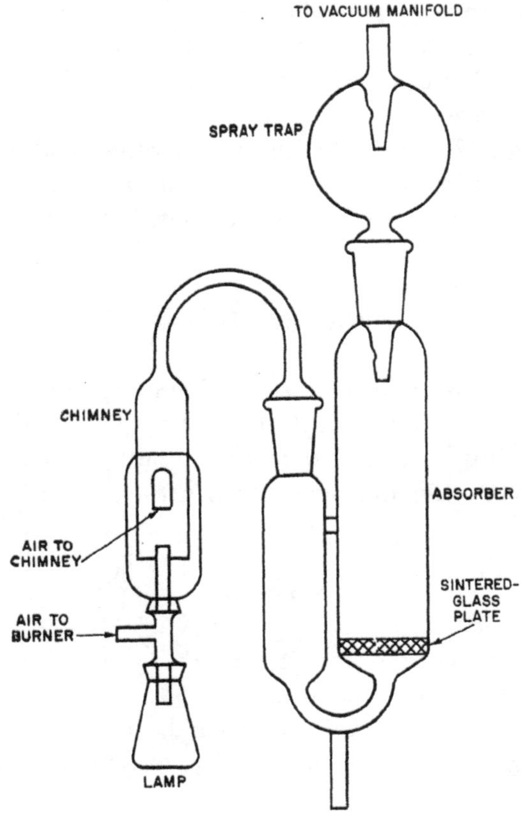

FIGURE 1. Diagrammatic Sketch of Combustion and Absorption Apparatus for Sulfur Determination

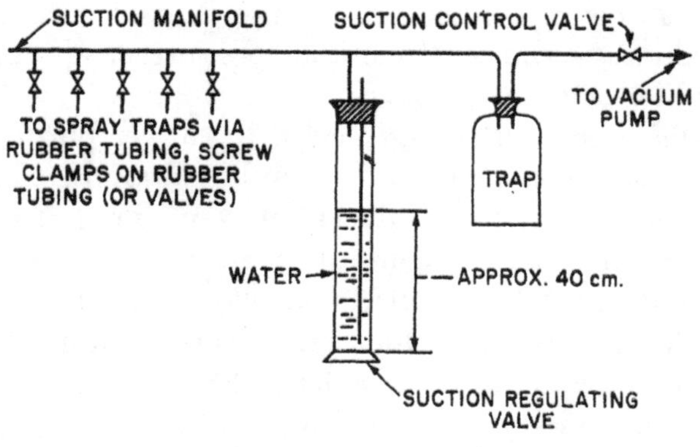

FIGURE 2. Suction System for Sulfur Determination

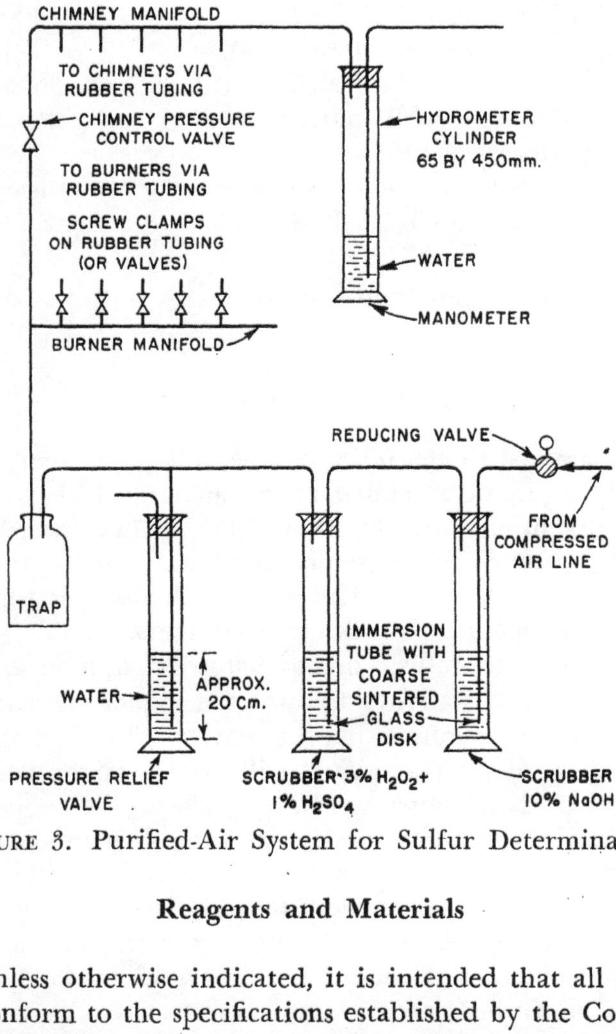

FIGURE 3. Purified-Air System for Sulfur Determination

Reagents and Materials

4. Unless otherwise indicated, it is intended that all reagents shall conform to the specifications established by the Committee on Analytical Reagents of the American Chemical Society, where such specifications are available. References to water shall be understood to mean distilled water.

(*a*) *Barium Chloride Solution* (*85 g per liter*)—Dissolve 100 g of $BaCl_2 \cdot 2H_2O$ in water and dilute to 1 liter.

(*b*) *Diluent*—The diluent used shall have a sulfur content less than 0.001 per cent, be completely miscible with the sample to be analyzed, and permit burning at a moderate rate without smoking. Normal heptane, *iso*octane, and absolute ethyl alcohol have been found suitable.

(*c*) *Hydrochloric Acid* (*sp gr 1.18*).

(*d*) *Hydrogen Peroxide* (*30 per cent*).

(*e*) *Hydrogen Peroxide* (*3 per cent*)—Dilute H_2O_2 (30 per cent) of low sulfur content with water. Store in a dark-colored glass-stoppered bottle.

(*f*) *Sodium Hydroxide Solution* (*100 g per liter*)—Dissolve 100 g of technical-grade NaOH pellets in distilled water and dilute to 1 liter.

(*g*) *Sulfuric Acid* (*1:16*)—Mix 60 ml of H_2SO_4 (sp gr 1.84) with 960 ml of water.

Preparation of Apparatus

5. (*a*) Place 300 to 400 ml of NaOH (100 g per liter) in the first scrubber (Figure 3) and the same amount of H_2O_2–H_2SO_4 solution [300 ml of water, 30 ml of H_2SO_4 (1:16), and 30 ml of H_2O_2 (30 per cent)] in the second scrubber. For apparatus in daily use, replace these solutions twice each week or whenever the volume becomes less than two thirds of the original.

(*b*) When the apparatus is first assembled, charge the absorber with 30 ± 2 ml of water, and adjust the individual valves between the vacuum manifold and the spray traps so that approximately 3 liters of air per minute will be drawn through each absorber when the chimney outlets are open to the atmosphere, while maintaining the pressure in the vacuum manifold at approximately 40 cm of water below atmospheric. When all adjustments have been made, remove the water from the absorbers.

(*c*) Introduce 30 ± 2 ml of H_2O_2 (3 per cent) into the larger bulb of each absorber. In addition to one absorber for each test sample to be burned, prepare an extra absorber for use as a blank. Attach the spray traps and chimneys and connect them to their respective manifolds by means of rubber tubing. Close the chimney openings by means of corks.

(*d*) With the burner-control valves closed, the valve to the vacuum regulator fully open, and the pressure in the vacuum manifold adjusted to approximately 40 cm of water below atmospheric, turn on the purified air. Adjust the chimney-manifold control valve so that, at the required flow through the absorbers, only a small stream of gas escapes at the pressure regulator, a small stream of air enters at the vacuum regulator, and the pressure in the chimney manifold is 1 to 2 cm of water. Minor adjustment of the vacuum regulator and vacuum-control valve may be necessary to achieve this condition.[1]

(*e*) Thread the required number of burners with four 15-cm lengths of cotton wicking by doubling two 30-cm lengths, inserting the looped ends into the top of the inner tube of the burner, and drawing the wicking through by means of a metal hook. Trim the wick as close as possible to the top of the burner with a pair of sharp scissors; then draw the wick down until the trimmed edge is flush with the top of the burner.[2] It is essential that thoroughly cleaned burners and new wicking are used for each test.

Procedure for Direct Combustion

6. (*a*) By means of a pipet, introduce into the flask of each burner an approximate quantity of sample as indicated in Table 1. Stopper the flasks with clean, numbered corks. Weigh each flask and its burner to 0.005 g.[3]

TABLE 1

Sulfur Content, per cent by weight	Sample Size g	Sample Size ml
0.5 and under	5 to 10	10
over 0.5	3 to 5	5

[1] It is convenient to balance the air-flow system by regulating the pressure in the vacuum manifold. This is done by raising or lowering the air-inlet tube in the vacuum regulator by sliding it in a rubber sleeve.

[2] Most types of samples burn with a luminous yellow flame, the size and shape of which is dependent on the air flow to the burner, the volatility of the material, the tightness of the fit of the wick in the burner tube, and the position of the top of the wick relative to the top of the burner. It is preferable that the latter two variables are fixed with relation to the first before burning is started so that the flame can be controlled by variation in rate of air flow. Highly volatile samples require a tight-fitting wick, the top of which may need to be several millimeters below the top of the burner. In extreme cases, the flask may have to be cooled in ice during the burning. Less-volatile materials require a more loosely fitting wick and may require warming. In any case, it is essential that the flame burns smoothly and symmetrically and without jets in the inner cone or smoke on the outer fringes. Materials that cannot be burned properly under these conditions shall be burned by blending as described in Section 7.

[3] While the stoppered flasks and prepared burners may all be weighed separately, it is usually more convenient to place each flask and its burner on the balance pan and obtain the combined weight in a single weighing.

(*b*) Handling each lamp individually, insert the burner in the flask. As soon as the sample has risen by capillary action to the top of the wick, connect the side tube of the burner to the burner manifold by means of rubber tubing. Light the burner with a sulfur-free flame (such as an alcohol lamp) and insert into the chimney, pinching off the connection between the chimney and the chimney manifold during the insertion, if the flame tends to be blown out. At the same time, adjust the air flow to the burner so that the flame is maintained at a point just below smoking and has a steady, symmetrical appearance. Continue in this manner until all lamps have been placed in the chimneys. Make any minor adjustment of the chimney-control valve necessary to maintain the required pressure [see Section 5 (*d*)]. During the burning, and particularly during the later stages when the flame becomes small, decrease the air supply to the burners in order to prevent extinction of the flames.

(*c*) When the burning of each sample is completed (as evidenced by the flame becoming small owing to depletion of the sample), remove the lamp from the chimney, extinguish the flame, shut off the air supply to the burner, and stopper the chimney opening. Immediately reweigh the lamp, burner, and numbered cork. When all combustions have been completed, turn off the air supply.

(*d*) Rinse the chimneys and spray traps with 30 to 40 ml of water, adding the rinsings to the absorbers, and proceed as directed in Section 8. When the sample burned contains tetraethyllead, use hot water to rinse the chimneys.

(*e*) *Blank*—Leave the chimney of the blank absorber [Section 5 (*c*)] stoppered and allow the purified-air stream to pass through that absorber until all samples started at one time have finished burning. Analyze the absorber liquid as directed in Section 8.

Procedure for Blending and Combustion

7. (*a*) Add 6 ml of sulfur-free diluent to each flask. Stopper the flasks with numbered corks and weigh to 0.005 g. By means of a pipet, introduce into the flask of each burner an approximate quantity of sample as indicated in Table 2, swirl to mix thoroughly, and reweigh.[3]

TABLE 2

Sulfur Content, per cent by weight	Sample Size g	ml
1 and under	3 to 5	5
over 1	2 to 3	3

(*b*) Insert the burner and burn as described in Section 6 (*b*). Remove each lamp from its chimney as the flame nears extinction and extinguish the flame. Add 2 ml of diluent, allowing the diluent to rinse down the walls of the flask. Burn the additional diluent, and repeat the addition of diluent and burning once more so that a total of 10 ml of diluent has been burned.

(*c*) After all lamps have completed burning, turn off the air supply. Rinse the chimneys and spray traps with 30 to 40 ml of water, adding the rinsings to the absorbers, and proceed as directed in Section 8.

(*d*) *Blank*—Prepare a separate flask containing 10 ml of diluent for use as a diluent blank and allow to burn while the samples are being burned. Analyze the absorber liquid as directed in Section 8.

Analysis of Absorbent

8. (*a*) Transfer the absorber liquid to a 400-ml beaker. Rinse the absorber thoroughly with water and add the rinsings to the beaker. Filter the solution to remove any foreign material, receiving the filtrate in a 400-ml beaker having a mark to indicate 75 ml. Add 2 ml of HCl, heat to boiling, and add 10 ml of $BaCl_2$ solution, either in a fine stream or dropwise. Stir the solution during the addition and for 2 minutes after. Cover the beaker with a fluted watch glass and continue boiling slowly until the solution has evaporated to a volume of approximately 75 ml, as indicated by the mark on the beaker. Remove the beaker from the hot plate (or other source of heat) and allow to cool 1 hour before filtering.

(*b*) Filter the supernatant liquid through a close-texture, ashless filter paper. Wash the precipitate with water, first by decantation and then on the filter paper, until free of chloride. Transfer the paper and precipitate to a suitable weighed crucible and dry at low heat until the moisture has evaporated. Char the paper completely without igniting it, and finally ignite at a bright red heat until the precipitate is burned white.[4] After ignition is complete, allow the crucible to cool to room temperature and weigh.

Calculation

9. Calculate the sulfur content as follows:

$$\text{Sulfur, per cent by weight} = \frac{(w - b) \times 13.73}{W}$$

where:

w = grams of $BaSO_4$ from sample,
b = grams of $BaSO_4$ from the corresponding blank,[5] and
W = grams of sample used.

Precision and Accuracy

10. (*a*) Results should not differ from the mean by more than the amounts listed in Table 3.

TABLE 3

Sulfur Content, per cent by weight	Repeatability, one operator and apparatus	Reproducibility, different operators and apparatus
0.002 to 0.25	0.001	0.002
over 0.25	0.5 per cent of mean	1 per cent of mean

[4] A satisfactory means of accomplishing these operations is to place the uncovered crucible containing the wet filter paper in a cold electric muffle furnace and turn on the current. Drying, charring, and ignition usually occur at the desired rate.

[5] The determination should be discarded if the blank correction used in the calculation exceeds 1.5 mg of $BaSO_4$. Frequently, impure reagents are the cause of this difficulty.

(*b*) Results should not differ from the true value by more than the following amounts:

Sulfur Content, per cent by weight	Accuracy
0.002 to 0.25	0.002
over 0.25	1 per cent of true value

APPENDIX

A1. *Apparatus*—The apparatus shall consist of the following:

(*a*) *Lamp*—A lamp of chemically resistant glass, consisting of a 25-ml Erlenmeyer flask and a burner that conforms to the dimensions shown in Figure 4. The burner consists of two concentric glass tubes, the external one having an arm, provided with standard-taper glass joints for connection with the flask and the chimney. The upper ends of both burner tubes shall be polished and shall have plane surfaces that are in the same horizontal plane. The burner shall have a 1-mm opening near its base to allow equalization of pressure between the chimney and the flask. When connected with the chimney, the lamp shall be held in position by rubber bands or metal springs stretched between glass hooks on the flask and chimney.

(*b*) *Chimney*—A chimney of chemically resistant glass conforming to the dimensions shown in Figure 4, provided with standard-taper glass joints for connection with the burner and absorber.

(*c*) *Absorber*—An absorber of chemically resistant glass conforming to the dimensions shown in Figure 4, provided with standard-taper glass joints for connection with the chimney and spray trap. A fritted disk with average pore diameter of 150 to 200 microns shall be sealed in the larger of two bulbs of the absorber. The fritted disk should be of such a porosity that, when 50 ml of water is placed in the absorber and air is passed through at the rate of 3.0 liters per minute in the forward direction, the pressure difference between the two sides of the absorber is between 15 and 23 cm of water and the air is dispersed uniformly.

(*d*) *Spray Trap*—A spray trap of chemically resistant glass conforming to the dimensions shown in Figure 4, provided with a standard-taper glass joint for connection with the absorber.

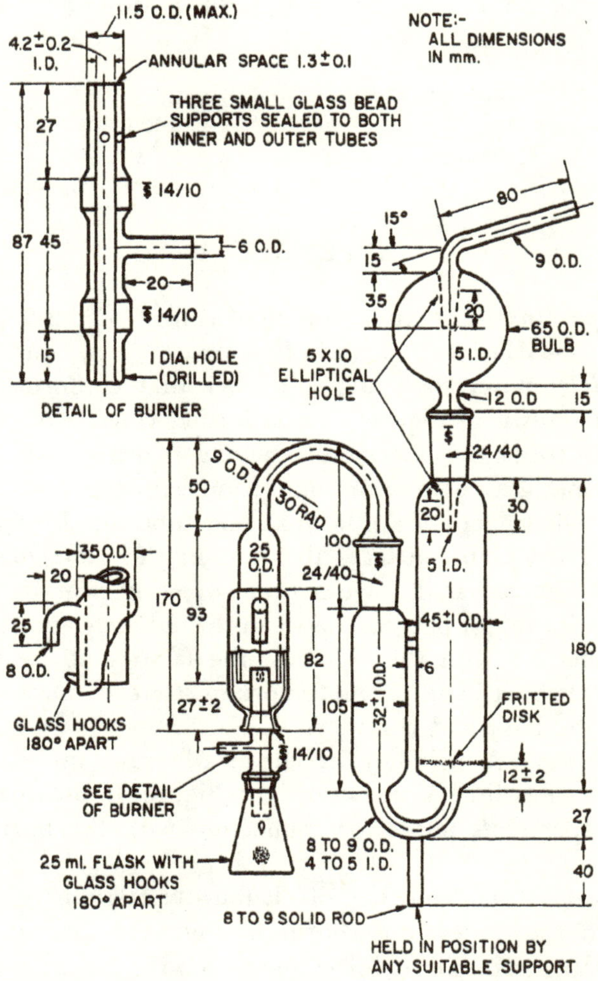

FIGURE 4. Detailed Drawing of Combustion and Absorption Apparatus for Sulfur Determination *

* In the case of those dimensions for which no specific tolerances are designated, the permissible variation is ± 10 per cent to the nearest millimeter, but in no case the deviation be greater than 5 mm.

DOCTOR TEST

This test is one of the oldest in common use for detecting the presence or absence of certain types of sulfur compounds in the sample. The method appears in Federal Specification VV–L–791 under the designation Method 5203. It is applicable to light petroleum distillates not heavier than kerosene.

A portion of the sample is mixed with one half its volume of "doctor solution," a small amount of sulfur added, and the mixture shaken vigorously. In the absence of certain types of sulfur compounds, notably mercaptans and sulfides, there will be no change in the color of any of the components of the mixture. In this case, the sample is considered *sweet,* and the test is reported as *passes.* If there is a change in color, the sample is considered *sour,* and the test is reported as *does not pass.* No attempt is made to measure the amount of sulfur present.

Apparatus

(1) *Sample Bottle*—A 4-oz bottle, fitted with a clean cork stopper.

Reagents

(1) *Doctor Solution*—A solution of litharge (PbO) dissolved in sodium hydroxide. To prepare, dissolve 125 g of sodium hydroxide in 1 liter of distilled water, add 60 g of litharge, and heat the mixture at 150 to 175°F. Stir the mixture at that temperature for 6 to 8 hours by means of compressed air. After cooling, add enough distilled water to make 1 liter of solution, allow the undissolved litharge to settle, and filter or decant the clear solution into a clean container. If the solution does not settle clear, it may be filtered through an asbestos mat. Protect the final solution from unnecessary exposure to air.

(2) *Sulfur*—Pure, dry flowers of sulfur, screened to 100 to 200 mesh. The reagent should be completely soluble in carbon di-

sulfide; if it is not, it can be made so by heating to 110°C for a few minutes.

Procedure

(1) Pour into a clean sample bottle 5 ml of the doctor solution and 10 ml of the sample under test. Stopper the bottle with a clean cork and shake for 15 seconds. Add a small amount of sulfur and shake again for 15 seconds. Allow the bottle to stand until the contents have separated cleanly.

It is important to avoid adding more sulfur than will just cover the interface. Some 20 to 25 mg is the proper quantity, which can be estimated with a little practice. If too much is added, any possible discoloration will be masked by the excess of sulfur.

(2) After settling, examine the layer of sulfur carefully. If the color of the sulfur is unchanged, report the sample as *passes*. If there is any evidence whatever of discoloration of the sulfur layer, or if there is a colored lace-like emulsion between the two liquid layers, report the sample as *does not pass*.

This examination must be made with extreme care. Sometimes the sulfur layer will be only flecked with spots of gray or black, and if there is any change in the color of either the sample or the doctor solution, these spots will be difficult to detect.

The relationship between the appearance of the sample and the type of sulfur is shown in Table 1.

TABLE 1

Appearance	Type of Sulfur
Black precipitate before addition of sulfur	Appreciable amounts of hydrogen sulfide
Black color of sulfur after shaking	Traces of hydrogen sulfide
Spots of discoloration in the sulfur layer, plus darkening of the sample	Hydrogen sulfide absent; mercaptans and/or elemental sulfur present
Clear yellow doctor solution; no discoloration of sulfur	Hydrogen sulfide and elemental sulfur absent; traces of mercaptans present

Strictly speaking, the test will not reject the sample on the basis of mercaptans only. The primary criterion for rejection is the appearance of the sulfur layer after shaking, and small amounts of mercaptans alone will not discolor the sulfur. However, samples which contain mercaptans will normally contain also small amounts of sulfur in other forms which will doscolor the layer at the interface.

When the sample contains appreciable amounts of hydrogen sulfide, a heavy black precipitate may form during the initial shaking and before the addition of the sulfur. If such a precipitate is noticed, the test may be stopped at that point and the sample reported as *does not pass.* However, if this observation is at all doubtful, the test should be continued.

DETERMINATION OF SULFUR IN PETROLEUM PRODUCTS HEAVIER THAN ILLUMINATING OILS

For specification purposes, the preferred method for the determination of sulfur in heavy oils is the same as that described in the current version of ASTM D–129, using the Parr oxygen bomb.

Under certain conditions, there are permissible alternatives which may be used. These alternates, and the reasons for the use of each, are as follows:

(1) The modified Dietert procedure was developed for the determination of sulfur in steel. This method is particularly useful when the final result must be available with a minimum of delay, since an answer can be obtained in less than 30 minutes. Extensive experiments in many laboratories have shown that the precision is at least as good as that of D–129. However, since the method has not been adopted by the ASTM, it should be used with caution.

(2) The Parr sodium peroxide fusion method. This is primarily useful for products containing large amounts of halogens, which will cause corrosion of the regular Parr oxygen bomb and will interfere with the Dietert procedure. This method is also useful in the examination of volatile liquids which contain large amounts of sulfur.

(3) The dilution-lamp method is applicable only to gas oils and distillate fuels, and should be used when it is necessary to determine only whether a sample does or does not meet the maximum specification requirements.

It should be clearly understood that the modified lamp procedure will yield results which are consistently low by about 10 per cent. This procedure should, therefore, be used only when it is necessary to know only whether a sample does or does not meet the maximum specification requirement. If the sample under test shows a sulfur content of less than 85 per cent of the permissible maximum when tested by this method, it may be assumed that the product is satisfactory, providing that the sample does not contain residues from acid-treated stock.

Procedure

The sulfur may be determined by any of the variations listed under L.I.C. 60.01. If purified air is used as the burning medium, the scrubber should be partly filled with 3 per cent hydrogen peroxide and the liquids titrated with standard sodium hydroxide at the end of the burning period. The charge burned consists of a 50–50 weight mixture of the sample and a kerosene of known sulfur content. This mixture should be prepared in a separate flask, agitated thoroughly, and a portion taken for the test. Two grams of this mixture should be burned, and the sulfur content found by the equations

$$\text{Percentage of sulfur in mixture (M)} = \frac{\text{Milliliters of NaOH}}{\text{Weight of mixture burned} \times 10}$$

and

$$\text{Percentage of sulfur in sample} = (M \times 2) - \text{Percentage of sulfur in kerosene}$$

Reporting

If the observed sulfur content is less than 85 per cent of the maximum permitted by the specification for the sample under test, the report should be worded as follows:

"Sulfur: Less than ..... per cent" (giving the maximum allowed by the specification).

If the observed content is 85 per cent or more of the permissible maximum, discard the results and determine the sulfur in some other manner.

Duplicate results obtained by this method should not differ from the mean by more than 4 per cent of the mean.

ASTM DESIGNATION: D 129–52

Scope

1. This method describes the procedure for the determination of sulfur in petroleum products, including lubricating oils containing additives, additive concentrates, and lubricating greases, that cannot be burned completely in a wick lamp. The method is applicable to any petroleum product sufficiently low in volatility that it can be weighed accurately in an open sample boat.

Outline of Method

2. (*a*) The sample is oxidized by combustion in a bomb containing oxygen under pressure. The sulfur, as sulfate in the bomb washings, is determined gravimetrically as barium sulfate.

(*b*) *Safety—Strict adherence to all of the provisions prescribed insures against explosive rupture of the bomb, or a blowout, provided the bomb is of proper design and construction and in good mechanical condition. It is desirable, however, that the bomb is enclosed in a shield of steel plate at least ½ in. thick, or equivalent protection provided against unforseeable contingencies.*

Apparatus and Materials

3. The apparatus and materials shall consist of the following:

(*a*) *Bomb,* having a capacity of not less than 300 ml, so constructed that it will not leak during the test and that quantitative recovery of the liquids from the bomb may be readily achieved. The inner surface of the bomb may be made of stainless steel or any other material that will not be affected by the combustion process or products. Materials used in the bomb assembly, such as the head gasket and lead-wire insulation, shall be resistant to heat and chemical action, and shall not undergo any reaction that will affect the sulfur content of the liquid in the bomb.

(*b*) *Sample Cup,* platinum, 24 mm in outside diameter at the bottom, 27 mm in outside diameter at the top, 12 mm in height outside, and weighing 10 to 11 g.

(*c*) *Firing Wire,* platinum, approximately No. 26 B. & S. gage.

(*d*) CAUTION—*The switch in the ignition circuit shall be of a type which remains open, except when held in closed position by the operator.*

(*e*) *Ignition Circuit,* capable of supplying sufficient current to ignite the cotton wicking or nylon thread without melting the wire.

(*f*) *Cotton Wicking or Nylon Sewing Thread,* white.

Reagents

4. Unless otherwise indicated, it is intended that all reagents shall conform to the specifications established by the Committee on Analytical Reagents of the American Chemical Society, where such specifications are available.

(*a*) *Barium Chloride Solution* (*85 g per liter*)—Dissolve 100 g of $BaCl_2 \cdot 2H_2O$ in distilled water and dilute to 1 liter.

(*b*) *Bromine Water* (*saturated*).

(*c*) *Hydrochloric Acid* (*sp. gr. 1.18*).

(*d*) *Oxygen,* free of combustible material and sulfur compounds, available at a pressure of 40 atmospheres.

(*e*) *Sodium Carbonate Solution* (*50 g per liter*)—Dissolve 135 g of $Na_2CO_3 \cdot 10H_2O$, or its equivalent, in distilled water and dilute to 1 liter.

(*f*) *White Oil,* U.S.P.

Procedure

5. (*a*) *Preparation of Bomb and Sample*—Cut a piece of firing wire, 100 mm in length. Coil the middle section (about 20 mm) and attach the free ends to the terminals. Arrange the coil so that it will be above and to one side of the sample cup. Insert between two loops of the coil a wisp of cotton or nylon thread of such length that one end will extend into the sample cup. Place about 5 ml of Na_2CO_3 solution in the bomb[1] and rotate

[1] After repeated use of the bomb for sulfur determinations, a film may be noticed on the inner surface. This dullness should be removed by periodic polishing of the bomb. A satisfactory method for doing this is to rotate the bomb in a lathe at about 300 rpm and polish the inside surface with grit No. 2/0, or equivalent paper,[3] coated with a light machine oil to prevent cutting, and then with a paste of grit-free chromic oxide and water. This procedure will remove all but very deep pits and put a high polish on the surface. Before using the bomb, it should be washed with soap and water to remove oil or paste left from the polishing operation.

Emery polishing paper grit No. 2/0 may be purchased from the Behr-

the bomb in such a manner that the interior surface is moistened by the solution. Introduce into the sample cup the quantities of sample and white oil [2] specified in Table 1, weighing the sample to the nearest 0.2 mg. (When white oil is used, stir the mixture with a short length of quartz rod and allow the rod to remain in the sample cup during the combustion.)

CAUTION—*Do not use more than 1 g total of sample and white oil or other low-sulfur combustible material.*

TABLE 1

Sulfur Content, per cent	Weight of Sample, grams	Weight of White Oil, grams
5 or under	0.6 to 0.8	0.0
over 5	0.3 to 0.4	0.3 to 0.4

(*b*) *Addition of Oxygen*—Place the sample cup in position and arrange the cotton wisp or nylon thread so that the end dips into the sample. Assemble the bomb and tighten the cover securely. CAUTION—*Do not add oxygen or ignite the sample if the bomb has been jarred, dropped, or tilted.* Admit oxygen slowly (to avoid blowing the oil from the cup) until a pressure is reached as indicated in Table 2.

TABLE 2

Capacity of Bomb, milliliters	Minimum Gage Pressure,[3] atmospheres	Maximum Gage Pressure,[3] atmospheres
300 to 350	38	40
350 to 400	35	37
400 to 450	30	32
450 to 500	27	29

Manning Co., Troy, N. Y. Chromic oxide may be purchased from J. T. Baker & Co., Phillipsburg, N. J.

[2] If the sample is not readily miscible with U.S.P. white oil, some other low-sulfur combustible diluent may be employed in place of the white oil. However, the combined weight of sample and nonvolatile diluent shall not exceed 1 g.

[3] The minimum pressures are specified to provide sufficient oxygen for complete combustion, and the *maximum pressures represent a safety requirement.*

(*c*) *Combustion*—Immerse the bomb in a cold distilled-water bath. Connect the terminals to the open electrical circuit. Close the circuit to ignite the sample. Remove the bomb from the bath after immersion for at least 10 minutes. Release the pressure at a slow, uniform rate so that the operation requires not less than 1 minute. Open the bomb and examine the contents. If traces of unburned oil or sooty deposits are found, discard the determination and thoroughly clean the bomb before again putting it in use.[1]

(*d*) *Collection of Sulfur Solution*—Rinse the interior of the bomb, the oil cup, and the inner surface of the bomb cover with a fine jet of distilled water, and collect the washings in a 600-ml beaker having a mark to indicate 75 ml. Remove any precipitate in the bomb by means of a rubber policeman. Wash the base of the terminals until the washings are neutral to a suitable indicator. (The volume of the washings is normally in excess of 300 ml.) Place the sample cup in a 50-ml beaker and add 2 ml of HCl and enough distilled water just to cover the cup. Heat the contents of the beaker to just below its boiling point for 3 or 4 minutes and add to the beaker containing the bomb washings. Wash the sample cup and the 50-ml beaker thoroughly with distilled water. Remove any precipitate in the cup by means of a rubber policeman. Add the washings from the cup and the 50-ml beaker, and the precipitate, if any, to the bomb washings in the 600-ml beaker. Do not filter any of the washings, since filtering would remove any sulfur present as insoluble material.

(*e*) *Determination of Sulfur*—To the combined washings add 10 ml of saturated bromine water and evaporate to approximately 200 ml on a hot plate or other source of heat. Adjust the heat to maintain slow boiling of the solution and add 10 ml of $BaCl_2$ solution, either in a fine stream or dropwise. Stir the solution during the addition and for 2 minutes after. Cover the beaker with a fluted watch glass and continue boiling slowly until the solution has evaporated to a volume of approximately 75 ml, as indicated by a mark on the beaker. Remove the beaker from the hot plate (or other source of heat) and allow to cool for 1 hour before filtering. Filter the supernatant liquid through an ashless, quantitative filter paper. Wash the precipitate with water, first by decantation and then on the filter, until free from chloride. Transfer the paper and precipitate to a suitable weighed crucible and dry[4] at low heat until the moisture has evaporated. Char the paper completely without igniting it, and finally ignite at a bright red heat until the residue is white in color. After ignition is complete, allow the crucible to cool to room temperature, and weigh.

(*f*) *Blank*—Make a blank determination whenever new reagents, white oil, or other low-sulfur combustible material are used. When running a blank on white oil, use 0.3 to 0.4 g and follow the normal procedure.

Calculation

6. Calculate the sulfur content of the sample as follows:

$$\text{Sulfur, per cent by weight} = \frac{(P - B)13.73}{W}$$

where:

P = grams of $BaSO_4$ obtained from sample
B = grams of $BaSO_4$ obtained from blank and
W = grams of sample used.

Precision

7. Results should not differ from the mean by more than shown in Table 3.

TABLE 3

Sulfur Content, per cent	Repeatability, same operator and apparatus	Reproducibility, different operators and apparatus
0 to 0.5	0.015	0.03
over 0.5 to 1.0	0.020	0.04
over 1.0 to 2.0[5]	0.025	0.05
over 2.0 to 3.0[5]	0.035	0.07
over 3.0 to 4.0	0.060	0.09
over 4.0 to 5.0	0.080	0.12
over 5.0	0.100	0.14

[4] A satisfactory means of drying, charring, and igniting the paper and precipitate is to place the crucible containing the wet filter paper in a cold electric muffle furnace and to turn on the current. Drying, charring, and ignition usually will occur at the desired rate.

[5] For lubricating greases in the range of 1.0 to 3.0 per cent sulfur, the repeatability and reproducibility are 0.05 and 0.10, respectively.

QUALITATIVE DETECTION OF SULFUR GASES IN INDUSTRIAL AROMATIC HYDROCARBONS

This method is based on ASTM D 853 and is normally performed concurrently with the distillation test described in L.I.C. No. 10.08. It is intended primarily for the examination of industrial aromatic hydrocarbons. The color of these is normally determined as described in the second section of L.I.C. No. 40.11. To be rated as water white, the color of the sample should not be darker than a solution containing 0.0030 g of potassium dichromate in 1 liter of distilled water.

Reagents

(1) *Lead Acetate Solution,* saturated.

(2) *Potassium Iodate Solution,* 10 per cent.

(3) *Starch Paper*—Dip strips of filter paper into starch solution and dry.

Procedure

(1) Assemble the apparatus for the distillation test as instructed in L.I.C. No. 10.08.

(2) Moisten a strip of the starch paper with the potassium iodate solution, and a strip of plain filter paper with the lead acetate. Hang both strips on the end of the condenser tube, placing them so that they are suspended in the upper part of the receiving cylinder and that drops of the condensate will pass between them without touching either one.

(3) At the end of the distillation test, if the lead acetate paper shows discoloration, hydrogen sulfide is present but sulfur dioxide is not. If the lead acetate paper shows no discoloration but the other develops a blue color, sulfur dioxide is present but hydrogen sulfide is not. If neither paper shows discoloration, neither gas is present.

DETECTION OF THIOPHENE IN BENZENE

This method will help detect the presence or absence of thiophene in benzene. It is similar to that described in ASTM D 931, but differs from it in that sulfuric acid is used as the catalyst, rather than ferric chloride. It is applicable to straight benzene or to the benzene separated from an alcohol containing this compound as a denaturant.

Using a porcelain spot plate or dish, 2 drops of concentrated sulfuric acid are mixed with a small amount of isatin, and 3 drops of the sample added. In the presence of thiophene, a green color will develop in a few minutes.

When the sample is a straight benzol, the test is made on the sample as received. When the sample is a denatured alcohol, a portion is distilled and the test performed on the first fraction of the distillate.

Apparatus

(1) *Distillation Apparatus*—The distilling flask should have a volume of about 1 liter. The receiver should have a volume of about 50 ml, and be fitted with a narrow neck. A volumetric flask or Babcock bottle may be used.

(2) *Spot Plate.*

Reagent

(1) *Isatin*—Melting point 200° to 201 °C.

Procedure

(1) When the sample is an alcohol, put 250 ml into the distillation flask and add 500 ml of distilled water. Assemble the apparatus and distill slowly until about 18 ml of distillate have been recovered, collecting the distillate in a flask with a narrow neck. Dilute the distillate to about 50 ml with distilled water and allow to stand until the layers have separated cleanly.

When the sample consists of a straight benzene, this distillation may be omitted.

(2) On a porcelain spot plate or dish, mix a small amount of isatin with 2 drops of concentrated sulfuric acid and add 3 drops of the benzene. In the presence of thiophene, a dark-green color will develop in a few minutes.

Sometimes a brown color will develop. To avoid confusion, it is suggested that two blanks are run simultaneously, one on thiophene-free benzene, and the other on benzene containing a small amount of added thiophene.

(3) If a green color develops, report the sample as *thiophene present.*

ASH AND WATER-SOLUBLE ASH CONTENT OF PETROLEUM OILS

This method is based on ASTM D 482, but the scope has been broadened to permit the determination of the water-soluble portion of the ash.

The method is applicable to all types of petroleum oils, either new or used. When the sample under test contains lead compounds or soaps of the alkali metals, the result does not express the true metallic content, since some of the metal will be lost by vaporization.

Apparatus

(1) *Ignition Dish* of about 100-ml capacity, made of platinum, porcelain, or silica. Whenever possible, it is strongly recommended that a platinum dish is used, since it is difficult to obtain a constant weight on dishes made of other materials.

CAUTION. A platinum dish *must not* be used if the sample contains lead. If lead compounds are present in the sample, during the combustion, the lead will be reduced to the metallic state and in that form will fuse readily with the platinum. Since used motor oils almost invariably contain lead, such samples should be ignited in a dish made of porcelain or silica.

(2) *Evaporating Dish* made of glass, of about 100-ml capacity.

(3) *Muffle Furnace* electrically heated and capable of being maintained at any desired temperature.

Care of Platinum Dishes

(1) Dirty dishes may be cleaned by adding a small amount of potassium acid sulfate, heating until the salt melts, and rotating the dish so that the melt will come into contact with all interior surfaces. After cooling, dissolve the melt in water, and finally rinse the dish with distilled water. If necessary, any remaining spots may be cleaned by rubbing with fine sand.

Under no circumstances should the dish be cleaned with hydrochloric or nitric acid. When these two acids are mixed, the mixture will liberate free chlorine which will attack the dish rapidly. If either acid is used for cleaning, and a salt of the other is present, chlorine will be liberated.

(2) While hot, the dish must not come into contact with any other metal and should, therefore, be handled only with tongs which are tipped with platinum. In addition, during the initial ignition, the dish must not be placed on a bare nichrome triangle. The supporting triangle should be either plated with platinum or covered with porcelain.

Procedure

(1) Heat the dish to redness, either in a furnace or over a gas flame. Cool the dish in a desiccator and weigh it to the nearest 0.1 mg.

(2) Introduce a suitable amount of sample into the dish, and weigh again. Record the difference as the weight of sample used for this test.

For samples containing ash in the range of 0.02 to 0.20 per cent, a 20-g sample is satisfactory. If the expected ash content is beyond this range, the sample weight may be increased or decreased, as necessary.

It is vitally important that the portion taken for the test is representative of the sample as a whole.

(3) Put the dish on an electric hot plate which is located in a hood. Heat gently by means of a small gas burner until the vapors can be ignited. Remove the burner and allow the oil to burn until the flame goes out.

If the sample contains moisture, there will be some foaming or spattering as the temperature approaches 212°F. In this case, stop the heating, add a few milliliters of absolute alcohol, ignite the alcohol vapors with the burner, and allow these vapors to burn without further heating. Continue this alcohol treatment until there is no further spattering. If necessary, remove the dish from the plate during this step.

Unless the sample is a very light distillate, some additional heat must be applied during the initial burning period. However, the flame must not be more than 1 or 2 in. high, to avoid mechanical loss of some of the ash. Proper burning conditions can be maintained by putting the dish on an electric hot plate; with well-controlled adjustment of heat, this operation will require no further attention on the part of the operator. Be sure to put a piece of asbestos or transite board on the plate when a platinum dish is used.

(4) When the initial burning has stopped, again heat the dish carefully with the gas burner until all volatile matter has been consumed. Then put the dish into a muffle furnace maintained at 1150°F (621°C) and leave it there until the carbonaceous matter has been burned and only the ash remains.

The treatment with the gas burner serves simply to remove any remaining volatile matter and thus prevent too violent combustion when the dish is placed in the furnace. The use of a gas burner for the final heating is not recommended; it is almost impossible to control the temperature accurately, and the flow of gases around the dish may remove some particles of the ash.

In some cases, the ash may fuse and entrap particles of carbon in the melt. It is almost impossible to remove the last traces of such carbon by heating alone. If this occurs, it is permissible to remove the dish, cool thoroughly, add a few milliliters of distilled water, and stir the mixture carefully with a glass rod. After the melt has been dissolved or dispersed, the dish should be put on a steam bath, the water evaporated, and ignition continued. This treatment may be repeated as necessary. If there is any doubt as to the purity of the distilled water, a blank test should be made on the same volume of water and a correction applied.

(5) Cool the dish in a desiccator and weigh again. Repeat the ignition until the weight is constant. Note and record the increase in weight of the dish, calculate the percentage, and report as .... per cent *ash*.

(6) If a determination of water-soluble ash is desired, put the dish into a beaker, add about 350 ml of distilled water, cover the beaker with a watch glass, and boil gently for about 15 minutes. Remove the dish from the water and wash any residue into the beaker with a fine jet of distilled water. Heat the beaker until the volume of the water is reduced to about 50 ml.

(7) Filter the liquid through coarse filter paper, collecting the filtrate in a tared glass evaporating dish. Rinse the beaker and paper with distilled water and filter the rinsings into the

dish. Put the dish on a steam bath and evaporate the contents to dryness.

(8) Remove the dish from the steam bath, wipe the outside with a clean cloth, and put it into an oven maintained at 221°F (105°C). At the end of 1 hour, remove the dish from the oven, cool in a desiccator, and weigh again. Repeat the drying until the weight is constant.

(9) Note and record the increase in weight of the glass dish, calculate to percentage, and report as .... per cent *water-soluble ash*.

The water-soluble ash is normally determined only when the sample contains small amounts of certain alkali soaps. The presence of alkali can be detected by adding 1 or 2 drops of phenolphthalein indicator after the ash has been determined; a pink color indicates the presence of alkaline salts.

This method is intended only as a rough indication of the presence of metals in the sample, and the result is not a true picture of the original metallic content. As previously stated, certain metals are volatile at test temperatures. The others will be present in the ash as some oxide. The exact composition of the oxide will depend to a large extent on the degree of heating. For a better determination of the true metallic content reference is made to the sulfated ash test, as described in ASTM D 874.

Dishes made of either porcelain or silica can be cleaned by adding nitric and hydrochloric acids and boiling *gently*. This cleaning should always be done between tests. If any residue is allowed to accumulate in the dish, it will be impossible to obtain a constant weight.

Certain specifications of the United States Government require that the ash content of the oil under test be "none." According to information obtained from the Navy Department, such oils may be considered satisfactory if the ash content is not more than 0.003 per cent, and nonabrasive in character.

CLOUD AND POUR POINTS OF PETROLEUM PRODUCTS (DILUTE POUR POINT)

The test recommended for determining the cloud and pour points of petroleum products is the same as that described in the current version of ASTM D 97, subject to the qualifications outlined in the next paragraphs.

Apparatus

The individual items of apparatus recommended are of the same general pattern as in ASTM D 97, but the specifications are more rigid. As an example, the ASTM specifies a jar measuring $4\frac{1}{2}$ to 5 in. in height, and an *internal* diameter of $1\frac{3}{16}$ to $1\frac{5}{16}$ in. We recommend a jar $4\frac{7}{8}$ to 5 in. in height, with an *external* diameter of $1\frac{9}{32}$ to $1\frac{5}{16}$ in. It has been definitely established that, in some cases, these differences will cause a variation in the final result.

Procedure

When tests are made at low temperatures in an atmosphere of high relative humidity, it is permissible to put a second ring gasket around the jar to minimize atmospheric condensation. If used, this second ring gasket should be placed so that when the jar is inserted into the jacket, the top surface of this second gasket will be level with the top of the jacket. This applies to both cloud and pour points.

Some specifications of the Federal government (MIL-L-6082A is an example) require a dilute pour point. The method for the dilute pour point is the same as the regular one, with the single exception that the test portion consists of a mixture containing 70 per cent by volume of the sample under test and 30 per cent of a suitable solvent. The solvent, in turn, consists of a mixture containing 20 per cent of xylene and 80 per cent of a petroleum naphtha conforming to TT-N-95a, with the additional requirement that no visible cloud or precipitate will form at a temperature of −60°F.

ASTM DESIGNATION: D 97–47

Definitions

1. (*a*) *Cloud Point*—The cloud point of a petroleum oil is the temperature at which paraffin wax or other solid substances begin to crystallize out or separate from solution when the oil is chilled under definite prescribed conditions.

(*b*) *Pour Point*—The pour point of a petroleum oil is the lowest temperature at which the oil will pour or flow when it is chilled without disturbance under definite prescribed conditions.

Scope

2. (*a*) The test for cloud point is intended for use only on oils which are transparent in layers $1\frac{1}{2}$ in. in thickness.

(*b*) The test for pour point is intended for use on any petroleum oil.

Apparatus

3. The following apparatus shown in Figure 1 shall be used:

(*a*) *Test Jar*—A test jar, A, of clear glass cylindrical form, flat bottom, approximately $1\frac{3}{16}$ to $1\frac{5}{16}$ in. in inside diameter and $4\frac{1}{2}$ to 5 in. in height. An ordinary 4-oz oil-sample bottle may be used if it meets these requirements, and no test jar is available.

(*b*) *Thermometer*—An ASTM cloud and pour test thermometer, graduated in either Centigrade or Fahrenheit degrees as specified, having a range of −38° to +50°C or −36° to +120°F or an ASTM low cloud and pour thermometer having a range of −80° to +20°C or −112° to +70°F, and conforming to the requirements for this thermometer as prescribed in the Standard Specifications for ASTM Thermometers.

(*c*) *Cork*—A cork, C, to fit the test jar, bored centrally to take the test thermometer.

(*d*) *Jacket*—A jacket, D, of glass or metal, watertight, of cylindrical form, flat bottom, about $4\frac{1}{2}$ in. in depth, with inside diameter $\frac{3}{8}$ to $\frac{1}{2}$ in. greater than the outside diameter of the test jar.

(*e*) *Disk*—A disk of cork or felt, E, $\frac{1}{4}$ in. in thickness and of the same diameter as the inside of the jacket.

(*f*) *Gasket*—A ring gasket, F, about $\frac{3}{16}$ in. in thickness, to fit snugly around the outside of the test jar and loosely inside the jacket. This gasket may be made of cork, felt or other suitable material, elastic enough to cling to the test jar and hard enough to hold its shape. The purpose of the ring gasket is to prevent the test jar from touching the jacket.

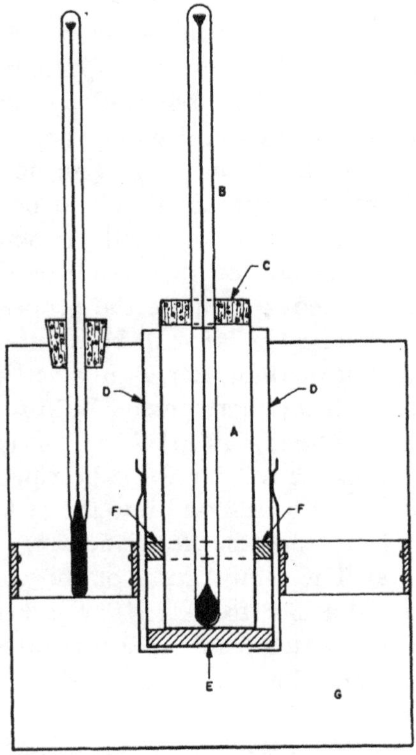

FIGURE 1. Apparatus for Cloud and Pour Test (as assembled for cloud test)

(*g*) *Bath*—A cooling bath, G, of a type suitable for obtaining the required temperatures. The size and shape of the bath are optional, but a support, suitable for holding the jacket firmly in a vertical position, is essential. For determinations of pour points below 50°F, two or more baths should be at hand. The required bath temperatures may be maintained by refrigeration if available, otherwise by suitable freezing mixtures.

The freezing mixtures commonly used are as follows:

	For Temperatures Down To
Ice and water	50°F
Crushed ice and sodium chloride	10°F
Crushed ice and calcium chloride crystals	−15°F
Solid carbon dioxide and acetone or gasoline [1]	−70°F

Procedure for Cloud Point

4. (*a*) The oil to be tested shall be brought to a temperature at least 25°F above the approximate cloud point. Moisture, if present, shall be removed by any suitable method, such as by filtration through dry filter paper until the oil is perfectly clear, but this filtration shall be made at a temperature at least 25°F above the approximate cloud point.

(*b*) The clear oil shall be poured into the test jar, A, to a height of not less than 2 nor more than 2¼ in. The test jar may be marked to indicate the proper level.

(*c*) The test jar shall be tightly closed by the cork, C, carrying the test thermometer, B, in a vertical position in the center of the jar with the thermometer bulb resting on the bottom of the jar.

(*d*) The disk, E, shall be placed in the bottom of the jacket, D, and the test jar with the ring gasket, F, 1 in. above the bottom shall be inserted into the jacket. The disk, jacket and inside of jacket shall be clean and dry.

(*e*) The temperature of the cooling bath, G, shall be maintained at 30° to 35°F. The jacket, containing the test jar, shall be supported firmly in a vertical position in the cooling bath so that not more than 1 in. of the jacket projects out of the cooling medium.

---

[1] This mixture may be made as follows: In a covered metal beaker chill a suitable amount of acetone or gasoline to 10°F, or lower, by means of an ice-salt mixture. Invert a cylinder of liquid carbon dioxide and draw off carefully into a chamois skin bag the desired amount of carbon dioxide, which through rapid evaporation will quickly become solid. Then add to the chilled acetone or gasoline enough of the solid carbon dioxide to give the desired temperature.

(*f*) At each test thermometer reading which is a multiple of 2°F, the test jar shall be removed from the jacket, quickly but without disturbing the oil, inspected for cloud, and replaced in the jacket. This complete operation shall require not more than 3 seconds. If the oil does not show a cloud when it has been cooled to 50°F, the test jar shall be placed in the jacket in a second bath maintained at a temperature of 0° to +5°F. If the oil does not show a cloud when it has been cooled to 20°F, the test jar shall be placed in the jacket in a third bath maintained at a temperature of −30° to −25°F.

(*g*) When such inspection first reveals a distinct cloudiness or haze in the oil at the bottom of the test jar, the reading of the test thermometer, corrected for error if necessary, shall be recorded as the cloud point.

Procedure for Pour Point

5. (*a*) The oil shall be poured into the test jar, A, to a height of not less than 2 nor more than 2¼ in. The jar may be marked to indicate the proper level. When necessary, the oil shall be heated in a water bath just sufficiently for pouring into the test jar.

(*b*) The test jar shall be tightly closed by the cork, C, carrying the test thermometer, B, in a vertical position in the center of the jar with the thermometer bulb immersed so that the beginning of the capillary shall be ⅛ in. below the surface of the oil.

(*c*) The oil shall be heated, without stirring, to a temperature of 115°F in a bath maintained at not higher than 118°F. The oil shall then be cooled to 90°F in air or in a water bath of approximately 77°F. Oils on which a pour point below −30°F is expected shall be heated as above with the high cloud- and pour-test thermometer in position, cooled to 60°F and the low cloud- and pour-test thermometer placed in position and the assembly placed in the jacket. Oils on which a pour point of above 90°F is expected shall be heated to 115°F or to a tempearture 15°F above the expected pour point, with the high cloud- and pour-test thermometer in position, and the test jar immediately introduced into the jacket.

(*d*) The disk, E, shall be placed in the bottom of the jacket, D, and the test jar, with the ring gasket, F, 1 in. above the bottom, shall be inserted into the jacket. The disk, gasket and inside of jacket shall be clean and dry.

(*e*) After the oil has cooled enough to allow the formation of paraffin-wax crystals, great care shall be taken not to disturb the mass of the oil nor to permit the thermometer to shift in the oil. Any disturbance of the spongy network of wax crystals will lead to low and fictitious results.

(*f*) The temperature of the cooling bath, G, shall be maintained at 30° to 35°F. The jacket, containing the test jar, shall be supported firmly in a vertical position in the cooling bath so that not more than 1 in. of the jacket projects out of the cooling medium.

(*g*) Beginning at a temperature 20°F before the expected pour point, at each test thermometer reading which is a multiple of 5°F, the test jar shall be removed from the jacket carefully and shall be tilted just enough to ascertain whether there is a movement of the oil in the test jar. The complete operation of removal and replacement shall require not more than 3 seconds. If the oil has not ceased to flow when its temperature has reached 50°F, the test jar shall be placed in the jacket in a second bath maintained at a temperature of 0° to +5°F. If the oil has not ceased to flow when its temperature has reached 20°F, the test jar shall be placed in the jacket in a third bath maintained at a temperature of −30° to −25°F. For determinations of very low pour points, additional baths should be maintained with successively lower temperature differences of about 30°F. In each case the test jar shall be transferred when the temperature of the oil reaches a point 50°F above the temperature of the new bath. At no time shall the cold test jar be placed directly in the cooling medium. As soon as the oil in the test jar does not flow when the jar is tilted, the test jar shall be held in a horizonal position for exactly 5 seconds, as noted by a stop watch or other accurate timing device, and observed carefully. If the oil shows any movement under these conditions, the test jar shall be immediately replaced in the jacket and a test for flow repeated at the next temperature, 5°F lower.

(*h*) The test shall be continued in this manner until a point is reached at which the oil in the test jar shows no movement when the test jar is held in a horizontal position for exactly 5 seconds. Certain lubricating oils tend to move as a whole and should be very closely observed. The reading of the test thermometer at this temperature, corrected for error if necessary, shall be recorded. The pour point shall be taken as the temperature 5°F above this solid point.

SPECIAL PROCEDURE FOR BLACK OILS, CYLINDER STOCKS AND NONDISTILLATE FUEL OILS

Special Procedure for Pour Point

6. (*a*) In those cases where it is known that a sample has been subjected to some temperature higher than 115°F during the preceding 24 hours or where the history of the sample in this respect is not known, the sample shall be held in the laboratory 24 hours before testing, unless three consecutive tests in accordance with Section 5 (*c*) of the same sample in the same test jar give check results. For these particular oils, results obtained by one or the other of the alternative procedures shall be called the upper (maximum) pour point.

(*b*) The lower (minimum) pour point shall be determined by heating a sample, while stirring, to 220°F. The oil shall then be poured into the test jar, cooled to 90°F as before, and the pour point determined as described in Section 5.

(*c*) Both the upper and lower pour points shall be reported.

Reproducibility of Results

7. Individual results of the pour test on the same oil in any one laboratory may vary by 5°F and in different laboratories by 10°F, although the average of three or more results in different laboratories should show a difference between averages no greater than 5°F. For oils tested by the special procedure described in Section 6, reproducibility of this order cannot be expected, as these oils show anomalous pour points depending on their previous thermal history.[2]

---

[2] It is a recognized property of these oils that the temperature to which they have been subjected before testing influences their pour points. Although the lower pour points, as determined by the special procedure, will show approximatly the reproducibility given above, yet the upper pour points will show greater variations depending on the previous thermal history of the oils. Further information on this subject is contained in *Proceedings,* Am. Soc. Testing Mats., Vol. 31, Part I, pp. 468 to 470 (1931), and Vol. 32, Part I, pp. 402 to 405 (1932).

DETECTING OXIDIZABLE MATTER IN SOLVENTS (PERMANGANATE TEST)

This method is intended primarily for the examination of solvents, such as acetone. It may be applied to other products, and shall be used whenever the Permanganate Test for reducing substance, is specified. The conditions of the test are identical with those of ASTM method D-268-49. A specific interpretation of the end point is made.

Apparatus

(1) *Constant-Temperature Bath*—Any convenient type may be used, providing that it can be held at a temperature of 25° ± 1 °C, and that it will accommodate one or more 250-ml Erlenmeyer flasks.

Reagents

(1) *Potassium Permanganate, 0.1 per cent by weight solution.* Dissolve 0.1 g of the chemically pure reagent in 100 ml of distilled water and allow the solution to stand in a dark place for 2 or 3 days. After settling, decant the solution into a clean amber bottle, discarding the first and the last few milliliters. The final solution should be protected from sunlight and should never be allowed to come into contact with rubber, filter paper, or other organic matter.

(2) *Mesityl Oxide.*

Procedure

Put 100 ml of the sample into a clean 250-ml Erlenmeyer flask and adjust the temperature to 25 °C. Add 1 ml of the permanganate solution, with slight stirring, and allow to stand in the dark. Bring the test solution into diffused light and observe the color relative to the standard described in the next paragraph. Record the time when the test solution fades to that of the standard. Record the time in minutes.

The difficulty of determining whether the pink color is retained can be overcome by the preparation of two samples as directed, but with three drops of mesityl oxide added to one. The sample with the mesityl oxide will fade rapidly, and its color can be used to determine the end point of the other sample.

Do not run the test longer than 2 hours. If the pink color is retained this long, report the value as 120+ minutes.

PERMANGANATE TIME TEST (IMPROVED VISUAL METHOD)

When contacted with a solution of potassium permanganate, an alcohol will be oxidized to the corresponding organic acid, with a decrease in the intensity of the color of the permanganate solution. This reaction is not specific for alcohols; other organic compounds, such as aldehydes, will behave in a similar fashion. However, the rate of reactivity for alcohols in general is lower than for the other organic compounds. Therefore, by conducting the test under carefully prescribed conditions and noting the time necessary for the decolorization of the permanganate, the purity of the alcohol under test can be estimated.

There are several methods in common use employing this principle. Since the details of procedure will affect the time for decolorization, no two different tests will give the same result. For comparability, it is, therefore, important that details are followed exactly. The method outlined here is similar to the one described by Weiss and Peters [*Anal. Chem.* **24,** 1658 (1952)], but contains certain modifications to improve the precision.

This method is intended for the examination of specification-grade alcohols. A portion of the sample under test is mixed with a solution of potassium permanganate and the resulting solution allowed to stand under carefully controlled conditions until the pink color of the permanganate has faded. The necessary time is noted and reported. Since the final solution will be colored slightly by the oxides of manganese, provision is made for comparing the color of the test solution with a set of standards.

Apparatus

(1) *Nessler Tubes,* 50-ml capacity.

(2) *Stirring Rods*—Each should be long enough to reach the bottom of a Nessler tube, and should have a flattened end.

(3) *Constant-Temperature Bath* capable of being held at 15° ± 0.1°C.

(4) *Daylight Fluorescent Lamp.*

Reagents

(1) *Potassium Permanganate Solution*—Dissolve 0.02 of the chemically pure reagent in 100 ml of distilled water. The solution should be prepared fresh at weekly intervals, kept in an amber glass-stoppered bottle, and never be allowed to come into contact with rubber, cork, filter paper, or other organic matter.

(2) *Standard Color Solutions*—Prepare separate solutions containing 50 g of cobaltous chloride ($CoCl_2{\cdot}6H_2O$) and 40 g uranium nitrate ($UO_2(NO_3)_2{\cdot}6H_2O$) per liter. Into separate Nessler tubes pipette the amounts of each solution shown in Table 1, dilute each to the 50-ml mark with distilled water, and mix thoroughly.

TABLE 1

Color Standard	Cobalt Solution milliliters	Uranium Solution milliliters
Before end point	5.0	6.5
Barbet color standard	5.0	7.0
Past end point	4.5	7.0

The second of these solutions represents the exact shade of color which is to be used as the end point. The other two represent the color immediately before the end point is reached, and after it has been passed.

Procedure

(1) Clean a 50-ml Nessler tube thoroughly with soap and water. Rinse the tube with either chromic or hydrochloric acid, then with distilled water, and finally with a portion of the sample to be tested. The acid rinse is particularly important if the tube has been used previously for this test.

(2) Put 50 ml of the sample under test into the clean Nessler tube and place the tube into the constant-temperature bath previously brought to a temperature of 15.0° ± 0.1°C. Insert a glass

stirring rod, and allow the tube to remain in the bath for 10 minutes.

The bath should be located so that it will be illuminated by soft, diffused light. Once the test has started, the tubes should never be exposed to strong sunlight.

(3) During this time, prepare a set of standards by pipetting the cobalt and uranium soluitons into separate Nessler tubes in the amounts listed in the table. Dilute each solution to the mark on the tube with distilled water, and mix with stirring rods.

(4) At the end of 10 minutes, add 2 ml of the permanganate solution to the sample under test, mix, and remove the stirring rod. Start a stopwatch at the beginning of the addition of the permanganate. Compare the color with the standard solutions without removing the tubes from the bath.

Illumination for the color comparisons should be from below the bath with a diffusion glass over a daylight fluorescent lamp. Eye color fatigue should be eliminated by examinations at short intervals rather than by prolonged staring. When not viewing the samples, the observer should look at neutral room surroundings rather than through a window or colored objects.

(5) When the color of the solution containing the sample under test is the same as that of the second of the three standard samples, stop the watch. Note the time in minutes, and report as *permanganate time.*

Experience has shown that better precision can be obtained if the test is stopped before the complete disappearance of the pink color.

Duplicate results by the same operator should not differ from the mean by more than 1.5 per cent of the mean. For different operators, the limit may be doubled.

It is vitally important that the following precautions are rigidly observed:

(*a*) At all times, the temperature of the sample–permanganate mixture must be 15.0° ± 0.1°C. An increase of 10°C may decrease the time by as much as 40 per cent.

(*b*) The mixture must be protected from direct sunlight, and the time for inspection should be kept to a minimum. Unless the approximate time is known at the start of the test, it is suggested that two portions of the sample under test are prepared simultaneously. One of these should be left undisturbed in the bath, and the other used for the preliminary observations. Once the approximate time is known, the final evaluation is based on the first portion.

(*c*) The reducing deposits from the permanganate will materially affect the reaction. The cylinder must, therefore, be cleaned with acid to remove oxides of manganese. However, the cleaning acid must also be completely removed. The cylinder is finally rinsed with the sample under test, since dilution of an alcohol with water will also affect the test.

(*d*) The sample must not be allowed to come into contact with organic matter of any kind. The sample should be taken in a bottle fitted with a glass stopper. If such bottles are not available, the usual cork stopper should be covered with clean metal foil.

Only alcohols of approximately the same strength should be compared. Dilution of a given sample with water will materially shorten the test time.

The possible impurities listed in Table 2, if present in the amounts indicated, were found to be sufficient to halve the time of a 50-minute alcohol.

TABLE 2

Impurity	Concentration, ppm
Acetaldehyde	3.500
Acrolein	0.015
Allyl Alcohol	0.005 to 0.100
Crude Fusel Oil	4.500

HOLDE HARD ASPHALT TEST

This method may be used on almost any type of product heavier than illuminating oils.

A portion of the sample is dissolved in forty times its volume of naphtha [1] and the solution allowed to stand in the dark for 24 hours. On the next day, the solution is filtered and the residue washed with naphtha [1] until the oily matter is removed. The residue, consisting of both asphalt and wax, is washed with hot absolute alcohol until all of the wax has been removed. The remaining asphaltic matter is dissolved in hot benzol, filtered, and the solution evaporated in a tared glass dish.

Procedure

(1) Weigh into an Erlenmeyer flask the proper amount of the sample under test. The proper weight of sample depends on the expected asphalt content, and may be found in Table 1.

TABLE 1

Expected Asphalt Content per cent	Weight of Sample grams	Volume of Solvent (Naphtha) milliliters
Below 0.1	10.0	440
0.1 to 0.5	5.0	220
0.5 to 2.5	2.5	110
Above 2.5	1.0	44

The weighing should be accurate to 0.1 per cent of the sample.

(2) Add the proper amount of solvent, as shown in the third column of Table 1, stopper the flask, and shake thoroughly.

[1] Originally a German product, known as "normal benzine," was specified for this test. Later it was found, however, that ASTM precipitation naphtha is perfectly suitable for this purpose.

Allow the mixture to stand for 24 hours in a dark place at 15° to 20°C.

(3) After standing for 24 hours, decant the solution through two filter papers (CS & S No. 589, white ribbon) and wash with more of the solvent until a portion of the filtrate leaves no oily residue on evaporation.

This operation must be carried out with care. As much as possible of the residue should be left in the flask. A transfer of much insoluble matter to the paper will result in a very slow filtration, and it will be almost impossible to remove all of the oily matter. However, the residue either on the paper or in the flask must never be allowed to become completely dry.

(4) When the oily matter has been removed, drain as much as possible of the remaining naphtha from the flask into the folded paper. Add 30 ml of absolute alcohol and heat the flask to gentle boiling on a hot plate. When most of the naphtha has drained through the paper, filter the alcoholic solution. Continue this washing with fresh portions of hot alcohol until the filtrate will leave no residue on evaporation.

The alcohol washing serves to remove the wax. Again, the residue either in the flask or paper should never be allowed to become completely dry, and as much as possible of the residue should be retained in the flask.

(5) When the alcoholic solution shows no trace of wax, put about 30 ml of benzol into the flask and heat to boiling. Remove the container from under the funnel and replace with a clean tared evaporating dish. Filter the hot benzol through the paper, collecting the washings in the dish. Repeat this extraction with fresh hot benzol until the washings are colorless.

(6) Transfer the dish to a steam bath and evaporate to dryness. Dry the dish in an oven for 1 hour at 105°C, cool in a desiccator, and reweigh. Note the increase in weight of the dish, and calculate the percentage.

DETERMINATION OF THE GUMMING PROPERTIES OF GASOLINE BY MEANS OF A COPPER DISH

This is one of the oldest methods in common use, but it has always been difficult for different laboratories to obtain concordant results. It is, therefore, imperative that the directions are followed closely, particularly regarding the assembly of apparatus.

This method is applicable to both motor and aviation gasolines. It must not be used for testing any product containing an additive which does not evaporate under the conditions of the test. The solvent oil used in motor gasoline is such an additive; the so-called aviation solvent oil is not.

Apparatus

(1) *Steam Bath*—The recommended bath is shown in Figure 1. This allows ten determinations to be made simultaneously, but a smaller bath of the same design may also be used. The openings for the dishes should be of 3⅜ in. diameter and should be ground true to circular form. This prevents the escape of steam around the sides of the dish, which is important. Each opening should be provided with a close-fitting cover, and any opening not in use should be covered. Heat may be provided by low-pressure steam or by actively boiling water; in the second case, the bath water should not come into contact with the outside of the dish. There should be a baffle to insure the even distribution of steam, an opening for a thermometer, and an excess-steam vent, at least 12 in. high.

(2) *Hood*—The bath should be mounted in a circular hood which can be tightly closed. To insure uniform distribution of air, a 1-in. hole should be bored into the bottom of the hood opposite the opening for each dish and centered 2 in. from the

side of the bath. Nipples of ¾ in. pipe should be placed in a vertical position in the openings, and ells threaded to the tops of the nipples. The nipples and ells should extend to approximately 3 in. above the top of the bath, and should be so turned that the emerging stream of air will be tangent to the bath. Suction should be provided to remove the gasoline vapors. If a fan is used, the motor driving it should be of the induction type and should be explosion-proof. A reproducible draft has been

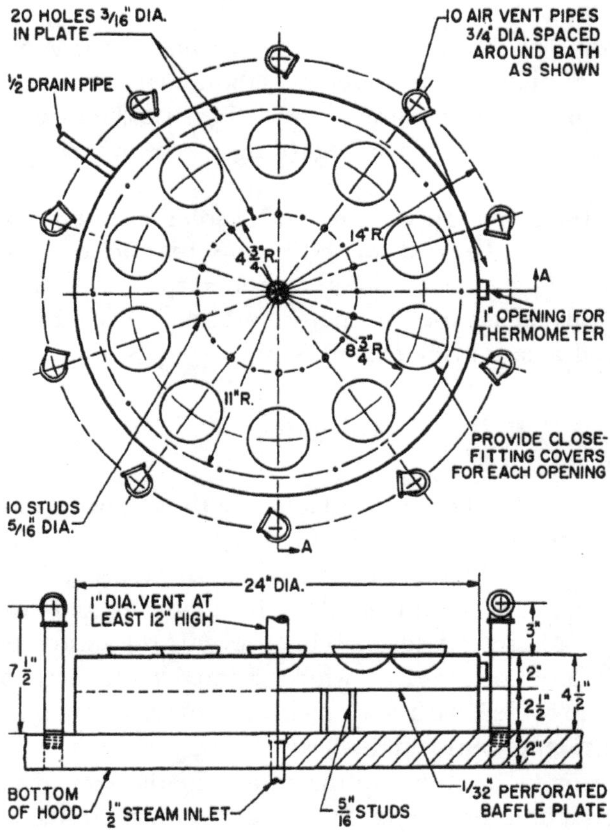

FIGURE 1. Steam Bath for Determination of Gum in Gasoline

obtained by installing a ¼-in. high-pressure air line in the stack of the hood and connecting a mercury manometer on the low-pressure side of the control valve.

A satisfactory assembly of bath and hood may be obtained from the Emil Greiner Co., 20–26 N. Moore Street, New York City, under their catalog number GR 2035. In this assembly, steam is to be used as the source of heat, and the hood is not fitted with a fan. It is recommended that this hood is modified slightly by inserting into the stack an inverted metal cone containing a center hole of 1¾ in. diameter. This will prevent any condensed liquid from dropping on the top of the bath, and permits close regulation of the draft.

(3) *Draft Gage*—In order to insure the evaporation of the sample within the specified time limits, the interior of the hood while in service should operate under a draft equivalent to 1 mm of water below atmospheric pressure. The draft may be measured by any type of gage of sufficient sensitivity. The differential Draft Gage, Type 1386, as sold by the McMaster Carr Supply Co., 640 Lake Street, Chicago, Ill. is satisfactory. If preferred, a device similar to the one shown in Figure 2 may be used. This consists of a small glass bulb, open at the top, and attached at the bottom to a slanting glass tube of approximately 3 mm internal diameter. This is mounted on a piece of plywood, which, in turn, is fastened to the hood just above the door. On the top of the board, there is a leveling device and behind the side arm, there is a scale which may be graduated in any convenient divisions. The top of the bulb is connected to the stack just below the cone by means of heavy-walled rubber tubing. To standardize the gage, the top of the bulb is attached to a manometer sufficiently sensitive to show a reading of 1 cm of water. The distance along the side arm may then be graduated in equivalents of 1 mm.

(4) *Copper Dish*—The dish should be a segment of a sphere measuring 3½ in. diameter by 1½ in. depth. The dish should be of spun copper, and the lip should be plain, not beaded.

(5) *Steel Wool*—For polishing the dishes, American grade, No. 000 steel wool should be used. A satisfactory product may be obtained from the American Steel Wool Co., 9 Desbrosses Street, New York City. As received, this is slightly oily and should be rinsed with naphtha and dried under vacuum before use.

(6) *Graduate* of 100-ml capacity.

The cleaning of the dish is probably the most important variable in the entire procedure. Any residue left from a previous determination should be thoroughly removed by solution in acetone after which the dish should be polished with the steel wool.

Polishing may be done by hand or a buffing wheel, but in all cases the surface must be entirely free of any discoloration. The ½ in. of surface next to the rim is usually the most difficult to clean. After polishing, the dish should be wiped with a clean dry cloth, dried in an electric oven at 100°C for ½ hour, and cooled in a desiccator.

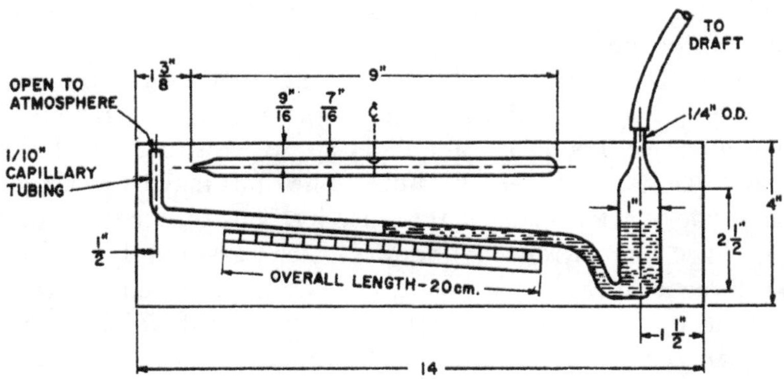

FIGURE 2. Draft Gage

Procedure

(1) The weight of the copper dish, cleaned and polished as prescribed in the preceding paragraph, should be determined to the nearest milligram, using a similar dish as a tare. The dish should then be placed in an opening of the steam bath, which should be at room temperature. Exactly 100 ml of the sample to be tested should be measured into a clean dry graduate and poured into the dish.

(2) Steam should be turned on slowly and maintained at a low rate for several minutes. This slow preliminary rate will prevent mechanical loss of gasoline through entrainment of liquid particles due to breaking bubbles which are formed initially with very volatile products. After the initial boiling subsides, steam should be turned on until a small quantity is seen escaping from the vent and the temperature of the bath is between 99.5° and 100°C.

(3) This temperature should be maintained until no more volatile substance remains in the dish. Normally this will require 2 to 3 hours for motor gasolines, and 40 to 60 minutes for avia-

tion gasolines. The dish is then removed from the bath, the outside wiped with a clean dry cloth, and dried in an oven for 30 minutes at 100°C. At the end of 30 minutes, the dish should be cooled in a desiccator and reweighed to the nearest milligram, using the same tare as for the original weighing.

(5) The increase in weight of the dish, expressed in milligrams; should be reported as *milligrams of gum, copper dish method.*

The results on the average gasoline should not differ by more than 10 per cent. Where the presence of high-boiling liquids requires more than 4 hours of evaporating time, the results are more difficult to check.

Precautions

(1) The surface of the dish *must be clean,* and should not be touched with bare fingers after the final polishing. Dishes which are slightly tarnished or which have a trace of gum left from a previous determination may give results two to three times as high as those properly prepared.

(2) The time of evaporation is important, and is affected by the rate of steam flow and of air circulation over the dishes. The steam flow should be just sufficient to keep the bath temperature constant, and the draft should have a pressure equivalent to 1 mm of water below that of the atmosphere.

DETERMINATION OF WATER IN PETROLEUM PRODUCTS BY THE KARL FISCHER REAGENT

This method is based on a number of similar tests appearing in the literature. It is applicable to all types of petroleum products and to many special solvents derived from petroleum. Mercaptans, hydrogen sulfide, strong alkalis, and tetraethyl lead interfere. When aldehydes and ketones are present, a special reagent must be used.

Apparatus

(1) *Titration Assembly*—All of the reagents used in this test are extremely sensitive to water and must, at all times, be protected from moisture in the atmosphere. The reagents are, therefore, stored in automatic delivery burettes, the openings of which are attached to drying tubes containing a satisfactory drying agent, such as Drierite. The titration is accomplished in a closed system completely protected from the air. A satisfactory assembly is shown in Figure 1.[1]

(2) *Magnetic Stirring Apparatus*—The set includes a rheostat, a magnetic stirrer, and a stirring bar. It is available from A. H. Thomas, W. Washington Square, Philadelphia, Pa., under their catalog number 9235-R. When ordering, a bar of 1½ in. length should be specified.

(3) *Galvanometer*—It is equipped with transformer, of the lamp and scale type, with a sensitivity of 0.02 microamperes per scale division. A suitable model may be obtained from Leeds and Northrup Company, 4901 Stenton Avenue, Philadelphia, Pa. under their catalog number 2420C. Accessory items include a dry cell and two resistors, one of 50 and the other of 4000 ohms, as shown in Figure 1.

(4) *Titrator*—Hafco titrator, as supplied by the Beckman Instrument Company, So. Pasadena, California.

---

[1] This may be obtained from Standard Inspection Supply (Enjay Co., Inc.).

Items (3) and (4) are alternatives and only one of the two is needed, the Hafco titrator being the more modern instrument.

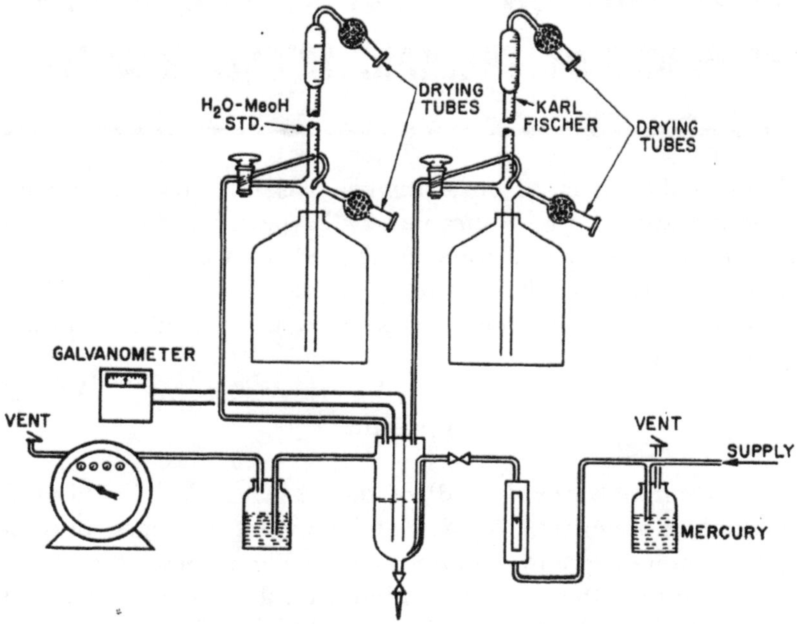

FIGURE 1. Titration Apparatus for Water in Liquid Products by the Karl Fischer Method

The Hafco titrator, like the galvanometer, measures a *dead stop* end point, but operates on the principle of the *magic eye*. On the top of the panel there are three lights, marked "Burette control," "Titrating," and "Stand-by." During the early stages of the test, the "Titrating" light is on, but the others are off. As the end point is neared, the "Titrating" light goes off and the "Stand-by" light will flicker. When the "Stand-by" light remains on for 30 seconds, the end point has been reached and the test is stopped.

As originally planned by the manufacturer, the complete assembly included an electrically controlled burette, so that the titration would be entirely automatic, and a reaction flask with built-in electrodes. These features were found to be unsatisfactory, and were eliminated. Since these accessories are not needed, any order on the Beckman Company should specify that only the Hafco titrator is wanted, not their complete assembly.

As received, the Hafco titrator can be plugged directly into any 110-volt 60-cycle line, so that the battery and resistors shown in Figure 1 may be omitted. A small adaptor, which can be made by any competent electrician, may be needed to attach the electrodes to the titrator. All other items shown in Figure 1 can be used without change.

Reagents

(1) *Methyl Alcohol*—Two types are required, one anhydrous and the other containing 1 mg of water per milliliter. Both of these are special reagents for this test and are available from any supply house.

(2) *Pyridine*—Grade 2A, refined, supplied by the Allied Chemical and Dye Corporation, Barrett Division, 40 Rector Street, New York City, is recommended. The ordinary grade of pyridine should not be used.

(3) *Sulfur Dioxide.*

(4) *Iodine.*

(5) *Solvent*—Mix 800 ml of anhydrous chloroform with 200 ml of absolute methanol. To remove any last traces of water, shake the mixture with silica gel, decant the clear liquid, and store in a tightly stoppered bottle.

(6) *Karl Fischer Reagent*—Two different types of reagent are required. The modified type is to be used only when aldehydes and ketones are present in the sample, and the regular type in all other cases.

(*a*) *Regular Reagent.* Dissolve 381 g of iodine in 1210 ml of pyridine in a large container fitted with a glass stopper, and add 3 liters of absolute methanol. Transfer 1 liter of this stock solution to a 2-liter flask, stopper the flask, and chill in an ice bath. While the solution is cooling, pass sulfur dioxide through sulfuric acid and collect the dry gas in a trap immersed in a dry-ice bath. Continue until 45 ml of liquid sulfur dioxide have collected in the trap. Slowly and with constant agitation, add the liquid sulfur dioxide to the cold stock solution. Shake the mixture until homogeneous and allow to stand for 1 or 2 days before use.

(*b*) *Modified Reagent.* Dissolve 84.7 g of iodine in 920 ml of pyridine. Chill the solution in an ice bath and when cold, add

45 ml of liquid sulfur dioxide. Shake until homogeneous and store in a tightly stoppered brown bottle for a day or two before use. As previously stated, this reagent is to be used only when the sample contains aldehydes and/or ketones.

Sampling

The sample must be drawn into a clean dry container fitted with either a metal screw cap or a cork stopper protected by metal foil. The sample container should be filled almost full and closed tightly. In the laboratory, when the sample under test is a petroleum product, this test should be the first one taken. The sample container should not be opened until everything is ready, and the test, including any checks if necessary, should be finished as soon as possible after the container has been opened.

Calibration

(1) Be sure that all the apparatus is clean and dry. Assemble the apparatus as schown in Figure 1, with the large burette attached to the top of the reaction flask, and the electrodes attached to either the galvanometer or the Hafco titrator, whichever is used. Fill the burette to the zero mark and put 75 ml of the special solvent into the separatory funnel.

(2) When everything is ready, allow the solvent to run into the reaction flask and immediately close the stopcock. Turn on the stirrer and the galvanometer (or titrator) and titrate the contents of the reaction flask with the reagent to a dead-stop end point. Note and record the volume of the reagent required.

The determination of the end point depends, of course, on the type of apparatus used. When using the galvanometer, add the reagent until the indicator registers zero for at least 30 seconds. When using the titrator, add the reagent until the "Titrating" light remains on for 30 seconds.

(3) When the water content of the solvent has been neutralized, pour into the separatory funnel about 100 mg of distilled water, weighed to the nearest 0.1 mg, using a Lunge weighing pipette or other similar device. Pour another 75 ml of the solvent into the funnel, allow the combined liquids to run into the reaction flask,

and repeat the titration, noting the volume of reagent required. Calculate the net volume of reagent used up for the water.

If preferred, the determination may be made by adding a known quantity of some salt containing a definite amount of water of crystallization, and titrating its water content. A suitable salt is hydrated sodium tartrate, $Na_2C_4H_4O_6 \cdot 2H_2O$. In this case, weigh some 200 mg of the salt into a small weighing bottle and transfer the entire bottle into the reaction flask. Add first about 100 ml of the absolute methyl alcohol, neutralize the water content, bring back the burette to the zero mark, and add the salt. No correction is thus necessary for any water in the solvent.

Whatever the procedure, it is extremely important that contamination from moisture in the atmosphere is avoided. Whenever any joint is opened to add a reagent, this should be done as quickly as possible and the joint closed immediately. When adding the sodium tartrate, store the filled weighing bottle in a desiccator until just before it is placed into the flask.

(4) Calculate the water equivalent of the Karl Fischer reagent by the equation:

$$F = \frac{S}{V}$$

where

S is the weight in milligrams of the sample of water poured into the flask.

V is the volume in milliliters of the reagent required for the titration.

F is the factor for the reagent, expressed as milligrams of water per milliliter of solution.

When the calibration is done by means of some hydrated solid, the numerator of the equation should be W × K, where W is the weight of the salt and K is the water content of the salt, expressed as a decimal. In the case of sodium tartrate, the value for K is 0.1566.

(5) Once the value for F has been determined, repeat the procedure, this time adding a known volume of the standard methanol solution to the clean, empty reaction flask, and determine accurately the water content of the solution.

The Karl Fischer reagent is not reliably stable, and the value for F will gradually change. For best results, the solution should be restandardized each day. When the test is performed at frequent intervals, it is sufficiently accurate for the standardization as described in the preceding paragraph and for correcting the value of F as needed. When done at infrequent intervals, the entire procedure should be repeated before a determination is made.

Procedure

(1) Assemble the apparatus as previously described. Pour about 100 ml of the solvent into the reaction flask. Select the proper type of Karl Fischer reagent, and insert the tip of the burette into the top of the flask.

(2) When everything is ready, rinse the separatory funnel with a portion of the sample under test and discard the rinsings. Then pour the proper amount of sample into the funnel, put the funnel into place, and close all openings tightly.

The size of the sample is shown in Table 1.

TABLE 1

Per Cent Water	Sample Size in Milliliters
0.0001 to 0.0500	100
0.0500 to 0.1000	50
0.1000 to 0.2500	20
0.2500 to 1.0000	5
Above 1	5.0 to 0.1

(3) With the stopcock on the funnel closed, neutralize the water content of the solvent by adding the reagent until an end point which is permanent for at least 30 seconds is reached. Do not record the volume, but refill the burette to the zero mark. Place the sample into the reaction flask, and repeat the titration to the same end point. Note and record the necessary volume of reagent.

When introducing the sample, lift the electrode assembly just enough to allow the sample to enter, and then reclose immediately.

(4) Calculate the water content of the sample by the equation:

$$W = \frac{V \times F \times 100}{S}$$

where

V is the volume in milliliters of the reagent required.
F is the strength of the reagent, as determined under CALIBRATION but expressed as GRAMS of water per milliliter of reagent.
S is the volume of sample in milliliters.
W is the water content of the sample, expressed as per cent by volume.

Precautions

It cannot be too strongly emphasized that precautions must be taken to prevent atmospheric moisture from affecting the analysis. The entire apparatus must be perfectly dry and, when once assembled, air should not be allowed to enter. The separatory funnel should be rinsed with a small portion of the sample and the rinsings either discarded or put into the reaction flask and neutralized.

It has been repeatedly shown that when the sample under test is a petroleum product and the sample container is only partly full, the final result will be affected by the volume of the vapor space above the liquid in the sample container. Under normal conditions, the greater this volume, the lower the final result will be. The sample container should, therefore, be filled almost full, and this test should be the first one taken.

When aldehydes and/or ketones are present, there are side reactions between these compounds and methanol, resulting in fictitiously high results. In most cases, errors from this source can be corrected by using the special reagent, which contains only a small amount of methanol. In the rare cases where the sample contains an appreciable amount, e.g., 50 per cent of methanol, some aldehydes or ketones, and a small amount of water, the methanol content of the special solvent should be decreased proportionately.

DETERMINATION OF WATER IN HYDROCARBON GASES BY THE KARL FISCHER REAGENT

The chemistry of the determination of water by the Karl Fischer reagent has been described in the preceding section (L.I.C. No. 150.04). With proper changes in operating details, this reagent can be used in a rapid routine method for determining water in various hydrocarbon streams. The following method has been used for the examination of isobutylene.

The procedure consists of passing a known volume of the sample through a scrubbing apparatus, similar to that shown in Figure 1. The water content of the scrubbing agent (methyl alcohol) is neutralized at the start, an excess of the reagent added, a known volume of sample passed through, and the water content determined in the regular manner.

In developing this method, the conditions described were found to be satisfactory. In particular, a scrubbing rate of 0.6 SCFH (cubic feet per hour under standard conditions) was found to give the best results. It is conceivable, however, that this rate might vary with the assembly of apparatus. It is therefore recommended that the entire routine listed under CALIBRATION is followed carefully.

The method is applicable to volatile hydrocarbons. Under the conditions described, duplicate results should not differ from the mean by more than 10 per cent of the mean. The accuracy is believed to be the same.

Apparatus

The complete assembly is shown in Figure 1. Reading from right to left, the individual items are as follows:

(1) *Pressure Relief Valve*—A bottle partly filled with mercury, with a T-joint connected to the source of sample and a vent to

the atmosphere. By regulating the level of the mercury, the flow of gas to the assembly can be controlled.

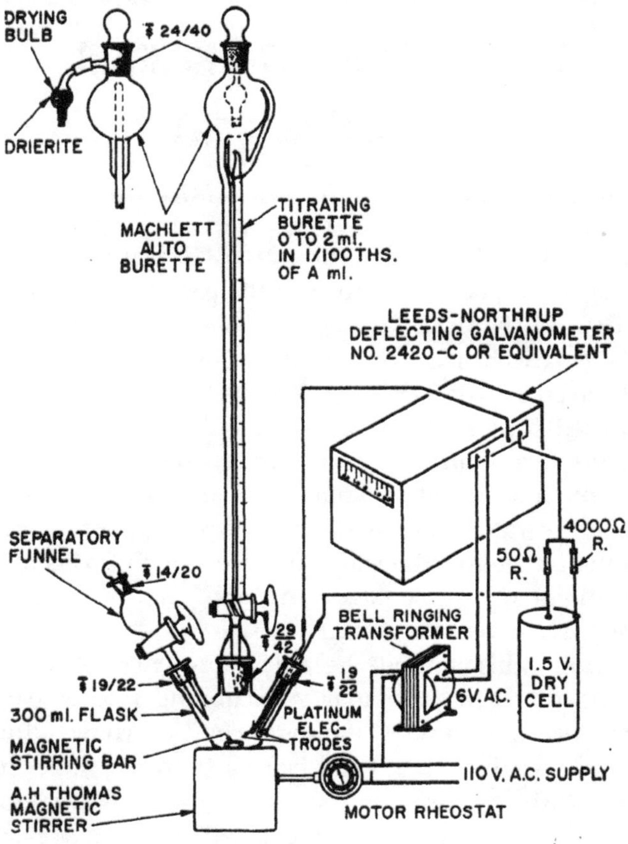

FIGURE 1. Titration Apparatus for Water in Gaseous Hydrocarbons by the Karl Fischer Method

(2) *Flowrator*—It covers the range of about 0.6 cu ft (17 liters) per hour. The line leading from the top of the Flowrator contains a stopcock, so that the flow of gas may be stopped as necessary. All connections should be made by means of Tygon tubing, and the lengths of tubing should be as short as possible.

(3) *Reactor*—A glass cylinder measuring about 3 × 7 in., with a tube containing a stopcock at the bottom, and equipped with a tight-fitting cover. The sample may be introduced through a tube blown into the side of the reactor, as shown, or through a

hole in the cover. In either case, the tip of the introduction tube should be fitted with a fritted glass disk of medium porosity. From the reactor, the sample passes through a scrubber, partly filled with caustic soda solution, to remove vapors of pyridine.

(4) *Wet-Test Meter*—Any convenient type, covering the range up to 5 cu ft per hour.

(5) *Galvanometer*—Sensitivity: 0.02 microamperes per scale division. A satisfactory item is available from Leeds and Northrup under their catalog number 2420C. The galvanometer may be activated by either a 6-volt battery or by regular house current, using a transformer. In either case, the galvanometer must be attached to a set of platinum electrodes which are partly immersed in the scrubbing liquid. A satisfactory set of electrodes is available from the Standard Inspection Supply.

The Hafco titrator, as supplied by the Beckman Instrument Company, South Pasadena, California, may be substituted for the galvanometer. This instrument, like the galvanometer, measures a *dead-stop* end point, but operates on the principle of the *magic eye*.

(6) *Burettes*—Automatic dispensing type, either of the kind shown in Figure 1 or the Machlett type as illustrated under L.I.C. No. 150.04 (preceding section). Two will be required, one for the Karl Fischer reagent and the other for the methanol. In either case, both must be connected to the reactor and the connecting lines should be as short as possible. All openings to the air must be protected by tubes filled with a moisture absorbent.

(7) *Weighing Pipette*—The Lunge or other convenient type (not shown in Figure 1).

Reagents

(1) *Karl Fischer Reagent*—Dissolve 762 g of resublimed iodine in 2.4 liters of pyridine. Shake until the iodine is dissolved and chill the mixture thoroughly in an ice bath. When cold, add 135 ml of liquid sulfur dioxide, swirl the mixture, and add 6 liters of dry methanol. Shake well and store in a tightly stoppered container. Determine the strength of this solution as described under CALIBRATION and dilute further with methanol to a water equivalent of about 0.25 mg of water per milliliter.

(2) *Methanol*—Anhydrous, available from any supply house.

TABLE 1. Optimum Scrubbing Rate for Accurate Water Content of Nitrogen Is 0.60 SCFH under Conditions Used

Gas Rate, SCFH, at Intervals during Total Run	Water content, parts per million: Karl Fischer at Intervals during Total Run	Water content, parts per million: Phosphorus Pentoxide at End of Total Run	Ratio of Karl Fischer Result to Phosphorus Pentoxide Result
0.46	53		
0.45	50		
0.39	58		
0.43 average	54 average	51	1.06
0.48	33		
0.44	35		
0.46 average	34 average	21	1.62
0.55	53		
0.51	49		
0.45	51		
0.43	43		
0.43	59		
0.43	71		
0.47 average	54 average	36	1.50
0.48	71		
0.44	60		
0.67	58		
0.46	55		
0.72	54		
0.55 average	60 average	58	1.03
0.61	34		
0.59	32		
0.60	30		
0.63	27		
0.63	30		
0.61 average	31 average	31	1.00

TABLE 1. (*Continued*)

	Water content, parts per million		
Gas Rate, SCFH, at Intervals during Total Run	Karl Fischer at Intervals during Total Run	Phosphorus Pentoxide at End of Total Run	Ratio of Karl Fischer Result to Phosphorus Pentoxide Result
0.55	52		
0.57	57		
0.63	63		
0.66	66		
0.64	64		
0.64	64		
0.65	65		
0.78	78		
0.64 average	64 average	66	0.97
0.66	35		
0.71	31		
0.65	36		
0.72	32		
0.69 average	34 average	37	0.92

Calibration

(1) Assemble the apparatus as shown in Figure 1. Put a 100-ml separatory funnel on top of the reactor, with the stem of the funnel protruding through a hole in the cap. Add enough of the dry methanol to cover the electrodes, disregarding the exact amount. Attach a cylinder of nitrogen to the opening normally used for the sample and put two towers, filled with a dehydrating agent, between the cylinder and the apparatus. Turn on the nitrogen, and purge the apparatus with the dry gas for about 5 minutes. This serves to remove any water vapor on the inside of the assembly.

(2) At the end of the purging period, titrate the contents of the reactor with the Karl Fischer reagent until any water is neutralized. Return the level of the liquid in the burette containing the reagent to the reference point, disregarding the amount added.

(3) Add through the funnel exactly 50 ml of the dry methanol. Titrate this portion with the Karl Fischer reagent, noting the amount added.

(4) Using a weighing pipette, weigh accurately about 40 mg of water into the separatory funnel. Add 50 ml of the dry methanol, and place the combined liquid into the reactor. Titrate again with the Karl Fischer reagent and note the necessary volume.

(5) Calculate the strength of the Karl Fischer reagent by the equation:

$$F = \frac{W}{V - B}$$

where

- V is the volume in milliliters of the reagent required for the water and alcohol.
- B is the volume for the alcohol alone.
- W is the weight of water added, in milligrams.
- F is the strength of the Karl Fischer reagent, expressed as milligrams of water per milliliter of solution.

(6) Using dry pyridine, adjust the value for F so that 1 ml of the Karl Fischer reagent will be the equivalent of 0.15 to 0.40 mg of water. Store the solution in one of the automatic burettes and repeat the standardization until consistent results are obtained.

(7) To 2 liters of the dry methanol, add enough distilled water so that the final content will be 0.50 to 0.75 mg/ml. Store this solution in the other automatic burette, and determine the actual water content by titration against the reagent.

(8) Once the strength of the solutions has been determined, the next step consists of finding the optimum rate of gas flow. To do this, disconnect the cylinder of nitrogen and replace it with a three-way stopcock. One opening of the stopcock should be attached to the assembly of apparatus and another to two drying towers arranged in series, each filled with phosphorus pentoxide. Attach a wet-test meter to the discharge end of the second tower and a source of compressed air to the third arm of the stopcock.

(9) Weigh the towers separately to the nearest 0.1 mg and place them in the system. Turn on the air and pass air through the

towers at a controlled rate for a given time. (Suggested conditions are 15 to 20 minutes and 0.6 cu ft per hour.) At the end of the time, stop the flow and note and record the total volume of air which has passed through the towers. Weigh the towers and record the weight of water absorbed.

(10) Put into the reactor a convenient volume (at least 5 ml) of the Karl Fischer reagent and pass the same volume of gas through the reactor. Stop the flow and back-titrate the contents of the reactor with the wet methanol. Calculate the weight of water absorbed by the Karl Fischer reagent and compare with that absorbed by the towers.

(11) Repeat this step, using various conditions of flow, until the amount of water found by the two methods is in good agreement. Record the optimum conditions of time and rate of flow and use these conditions in all future testing.

Typical values found during the work in establishing this method are listed in Table 1. These conditions may change according to the location. Optimum conditions must, therefore, be found by experiment.

The use of compressed air has been specified in determining the optimum rate of gas flow. If air is taken from the same source and passed in succession through the drying towers and the scrubbing solution, the moisture content of the two portions should be the same. If preferred, nitrogen may be used as the gas instead of compressed air. In this case, the nitrogen should be humidified by passing through a dilute solution of sulfuric acid which is in a container immersed in a constant-temperature bath.

Procedure

(1) With the apparatus completely assembled, attach the inlet end to the source of sample. Put into the reactor enough methanol so that the bottoms of the electrodes will be immersed. Adjust the flow of sample to the optimum rate, as found under CALIBRATION, and purge the apparatus with the sample for 5 minutes.

(2) Decrease the flow of sample somewhat and neutralize the water in the reactor, disregarding the exact amount of the reagent added. Stop the flow of sample and read the wet-test meter.

(3) Return the liquid level of the reagent to the reference mark in the burette and add a measured amount of the reagent to the reactor. Record the amount added.

(4) Open the stopcock and allow the sample to pass through the liquid for the required time at the proper rate. Stop the flow of sample and note the volume taken, the temperature of the liquid in the wet-test meter, and the barometric pressure.

(5) Determine the amount of excess Karl Fischer reagent by back-titrating with the standard methanol solution.

Calculations

(1) Calculate the weight of gas by the equation:

$$G = \frac{V \times 28{,}317 \times D \times P \times 273}{T \times 760}$$

where

V is the volume of sample, in cubic feet, which has passed through the reactor (the value 28,317 is the number of milliliters in a cubic foot).

D is the density of the sample under test, in grams per milliliter.

P is the observed barometric pressure, in millimeters of mercury.

T is the temperature of the water in the wet-test meter in °C.

G is the weight of the tested sample, in grams.

(2) Calculate the water content of the sample by the equation:

$$W = \frac{(K \times F) - (M \times F') \times 1000}{G}$$

where

K is the volume in ml of the Karl Fischer reagent used in the titration.

F is the water equivalent of this solution, in milligrams per milliliter.

M is the volume in milliliters of the methanol used in the back-titration.

F′ is the water equivalent of the methanol solution, in milligrams per milliliter.

W is the water content of the sample under test, expressed in parts per million.

The Karl Fischer reagent can be standardized either by the method outlined here, using a weighed amount of water, or by the use of a salt containing a definite amount of water of hydration, as described in L.I.C. No. 150.04. In non-air-conditioned areas, the second method may give serious errors because of the condensation of moisture on the surface of the weighing bottle.

Both the Karl Fischer reagent and the methanol are extremely sensitive to moisture in the air. All openings to the automatic burettes containing either of these liquids must be protected by drying tubes, filled with some moisture absorbent, and the contents of the tubes must be changed at frequent intervals.

Although some previous work has been done on determining water in hydrocarbon gases by the Karl Fischer reagent, the results were, in general, erratic. It is believed that the variations were caused largely by the rate of scrubbing of the gas and the standardization of the reagent. Both of these subjects have been amply covered by the procedure outlined here.

ACIDITY OF INDUSTRIAL AROMATIC HYDROCARBONS

This method is based on ASTM D–847. It is suitable for the detection and quantitative determination of acidity in industrial aromatic hydrocarbons.

Definitions

Acidity—Acidity is the number of milligrams of sodium hydroxide consumed when 100 ml of the sample are treated under the given conditions.

Acid Reaction—Acid reaction is producing the acid color of the indicator used under the conditions of this method.

Alkaline or Basic Reaction—This is producing the alkaline color of the indicator used under the conditions of this method.

Apparatus

(1) *Graduate*—A 100-ml graduated cylinder will be needed.

(2) *Tincture Bottle*—A 500-ml glass-stoppered bottle is used.

(3) *Burette*—The burette should have a capacity of 10 ml, and be graduated in 0.05-ml subdivisions.

Reagents

(1) *Neutral Distilled Water*—In a large flat-bottomed flask, 1 to 2 liters of distilled water are boiled vigorously for 30 minutes. A 200-ml flask is rinsed with this water, and a 100-ml portion put into the smaller flask. This portion is reheated to boiling and titrated immediately with 0.01 N NaOH, taking care that the temperature does not fall below 80°C during the titration. If more than 1 drop of the NaOH is required, sufficient alkali to neutralize the remaining portion is added to the original sample, and the titration repeated. Further adjustments are made if

necessary until the blank titration of a 100-ml portion of the distilled water is not more than 1 drop of 0.01 N NaOH. During this time, the main portion of the water is kept at or just below the boiling point. When the acidity has been properly adjusted, the water is allowed to cool somewhat, and the flask fitted with a soda-lime guard tube.

(2) *Indicator*—Exactly 0.5 g of phenolphthalein is dissolved in 100 ml of chemically pure 95 per cent ethyl alcohol, and the solution titrated with the 0.01 N NaOH to a faint-pink end point. The pink color is removed by the addition of 1 or 2 drops of 0.01 N acid.

(3) *Standard 0.1 N Sodium Hydroxide Solution*—A convenient quantity of chemically pure sodium hydroxide is dissolved in an equal weight of distilled water and the resulting solution stored in a tightly stoppered container. After standing for several days, 5.2 ml of the clear supernatant liquid is decanted off and diluted to exactly 1 liter with recently boiled and cooled distilled water. The solution is standardized with benzoic acid or acid potassium phthalate, and the strength adjusted to exactly 0.1 N. The final solution is stored in a chemically resistant bottle and protected by a soda-lime tube.

(4) *Standard 0.01 N Sodium Hydroxide*—In a similar manner, a 0.01 N solution of NaOH is prepared and standardized, using 0.5 ml of the solution.

(5) *Standard 0.01 N Acid*—Approximately 0.3 ml of concentrated sulfuric acid is diluted to exactly 1 liter with distilled water. The resulting solution may be standardized by titration against either of the standard alkali solutions.

Procedure

(1) Measure 100 ml of the sample, 100 ml of the neutralized distilled water, and 2 drops of the indicator into the tincture bottle. Stopper the bottle and shake thoroughly. Without separating the layers, titrate the mixture with the standard 0.1 N alkali to the first persistent shade of pink, swirling the bottle during the titration.

(2) If more than 2 drops of the 0.1 N solution are required, the results are discarded and the test repeated, taking stringent precautions to avoid contamination from the apparatus and at-

mosphere. This time, the graduate and the bottle are rinsed with the neutral distilled water. Then 100 ml of the sample, 100 ml of neutral distilled water, and 2 drops of indicator are introduced into the bottle, which is then stoppered and shaken for 10 seconds. The temperature of the liquids is brought to 60° to 65°F. One drop of the NaOH is added, and the bottle stoppered and shaken vigorously for 10 seconds. This is repeated, adding the alkali, one drop at a time, and shaking for 10 seconds after each addition, until a sharp pink end point is obtained. The procedure is repeated on a blank consisting of 100 ml of the neutral distilled water and 2 drops of the indicator.

Unless otherwise dictated by the specifications, the test results are interpreted as follows:

(1) A sample is said to contain no free acid (show no evidence of acidity) if 2 drops or less of 0.1 N NaOH produce a persistent pink end point in the test bottle.

(2) When more than 2 drops of 0.1 N NaOH are required to produce a persistent pink end point, the acidity is reported in terms of milligrams of NaOH per 100 ml of sample, and is calculated by the equation:

$$A = 4 \times (T - B)$$

where

T is the volume of 0.1 N NaOH required for the sample.
B is the volume for the blank.
A is the acidity of the sample, expressed in milligrams of NaOH per 100 ml of sample.

Precautions

Tests for acidity are not applicable in the presence of contaminating acidic or alkaline gases, soaps, salts, or other compounds derived from the atmosphere or apparatus.

The sample container, apparatus, water, indicator, etc., used in the test should be so chosen that they themselves do not affect the result appreciably. Since new corks used in sample bottles are frequently bleached with oxalic acid, it is advisable to rinse them thoroughly and check them for neutrality with the indicator used in the test. Glassware should be made of Pyrex or its equivalent and rinsed with neutral distilled water before use.

The test is performed in a room whose atmosphere is not contaminated by carbon dioxide, ammonia, or other interfering substances.

The distilled water used in the test should not alter the composition of the sample nor otherwise interfere with the test. Although inert impurities may often be neglected, care must be taken to correct for impurities or to eliminate them entirely if they are likely to interfere. When the distilled water does not show an acid or alkaline reaction, it may be used without further adjustment to neutrality. However, if it shows either reaction, it should be brought to a persistent pink end point before use by titration with standard 0.01 N acid or alkali, as necessary.

L.I.C. No. 160.50A

DETERMINATION OF THE UNSAPONIFIABLE MATTER IN, AND THE NEUTRALIZATION NUMBER OF, NAPHTHENIC ACIDS

This method was developed to replace the Spitz-Honig procedure which yields low results if the unsaponifiable matter is volatile. The procedure is based on a similar one, appearing in the literature, and has been modified to its present form as a result of cooperative work by several laboratories. It is suitable primarily for the examination of naphthenic acids.

Reagents

(1) *Standard 0.5 N KOH Solution*—Approximately 30 g of chemically pure potassium hydroxide, purified by alcohol, is dissolved in 1 liter of either water or alcohol. The resulting solution is stored in a bottle made of chemically resistant glass, protected by a soda-lime guard tube, and standardized against Bureau of Standards grade of benzoic acid or acid potassium phthalate.

If standardized against benzoic acid, approximately 2.5 g of the reagent is weighed accurately into an Erlenmeyer flask, 50 ml of alcohol added, and the solution titrated cold, using phenolphthalein. A blank determination is made on the same volume of alcohol, and the observed volume subtracted from that used for the sample. If standardized against acid potassium phthalate, approximately 4 g of the reagent is weighed accurately into a flask, dissolved in 100 ml of distilled water, and titrated cold, using phenolphthalein.

(2) *Alcohol*—This may be either 95 per cent ethyl, or 99 per cent isopropyl alcohol, as preferred. In either case, it should be neutralized just before use.

(3) *Phenolphthalein*—One gram of the indicator is dissolved

in 100 ml of alcohol, and the resulting solution neutralized, preferably with 0.1 N KOH solution.

(4) *Methyl Orange*—One gram of the solid is dissolved in 1 liter of distilled water.

(5) *Petroleum Ether*—This should have a boiling range of 35° to 60°C and be neutral to phenolphthalein. A satisfactory solvent is available from the Baker Chemical Co., Phillipsburg, N.J.

Procedure

(1) from 2 to 3 g of sample, accurate to the nearest milligram, is weighed into a clean dry flask, 50 ml of neutralized alcohol added, and the contents of the flask heated to boiling. The mixture is titrated with the 0.5 N KOH solution to a sharp pink end point, agitating the flask during the titration and reheating if necessary. The volume of KOH is recorded.

(2) An additional 0.5 ml of the KOH solution is added, and the contents of the flask transferred quantitatively to a 250-ml separatory funnel. Either alcohol or water, as needed, is poured into the funnel in the amount necessary to insure an alcoholic strength of 60 $\pm$ 5 per cent. Then 25 ml of petroleum ether is added, and the funnel stoppered and shaken vigorously. After settling, the alcoholic solution is drawn off into a second funnel and the ether washing repeated. If necessary, this washing should be repeated until all the unsaponifiable matter is removed from the alcoholic solution.

Normally, two washings are sufficient, but if the sample has a neutralization number of less than 190, more may be required. If more than two washings are needed, the petroleum ether should be collected, washed with alcohol, and the alcohol washings combined with the main portion.

(3) The alcoholic solution is drawn off into a clean beaker and heated on a hot plate or steam bath until all the alcohol has been evaporated. The aqueous solution is transferred to a clean separatory funnel, and 15 ml of dilute HCl (1:4) and 25 ml of petroleum ether added. The funnel is stoppered, shaken vigorously, and then put aside to allow the contents to settle. When separation is complete, the acid layer is drawn off, and the petroleum ether layer washed with distilled water until the water is neutral to methyl orange, discarding all the water washings.

(4) The petroleum ether layer is filtered through two pieces of filter paper, collecting the filtrate in a clean, dry tared flask. The flask is placed on a steam bath and heated until the ether is removed. The outside is then wiped with a clean, dry cloth, and the flask cooled in a desiccator and reweighed, recording the weight of the acids.

(5) The neutralization number of the regenerated acids is determined in the same manner as outlined under (1), recording the necessary volume of KOH.

Calculations

(1) The neutralization number of the acids may be found by the equation:

$$N = \left(\frac{V \times F}{W}\right)$$

where

- V is the volume of KOH used in the titration.
- F is the strength of the solution in milligrams of KOH per milliliter.
- W is the weight of sample.
- N is the neutralization number, expressed in milligrams of KOH per gram of sample.

(2) The percentage of unsaponifiable matter in the sample can be found by the equation:

$$U = 100 - \left(\frac{NO \times 100}{NR}\right)$$

where

- NO is the neutralization number of the sample as received, or original acids [PROCEDURE (1)].
- NR is the neutralization number of the regenerated acids [see (5) under PROCEDURE].
- U is the percentage of unsaponifiable matter in the sample.

ANILINE POINT

The method recommended for the determination of aniline point is the same as that described in the current version of ASTM D 611, with the exceptions outlined in the next paragraphs.

(a) The U-tube apparatus shown in Figure 1 is a permissible alternate for either of the assemblies illustrated in D 611. It may be used for the examination of all products on which an aniline point is normally required. Existing items of apparatus are not to be discarded and may continue to be used if desired.

(b) Since aniline is extremely toxic, the use of a special apparatus is recommended for the distillation of this reagent for the following two reasons:

(1) The exposure of the operator to liquid aniline and its vapors is kept to a minimum.

(2) Since the receiver is made of red glass, the aniline need not be freshly distilled. Tests have shown that distilled aniline can be stored in the receiver for 5 days without affecting the final result.

(c) ASTM D–611 states flatly that the aniline must be purified before use. Repeated experiments have shown that the distillation is not always necessary; it is now possible to buy a special grade of aniline (Baker's purified is satisfactory) which will meet the normal requirements for purity. It is, therefore, suggested that aniline of this quality is obtained, and that a blank test is made on *n*-heptane *each time* a series of tests is started. As long as the observed result is within the permissible limits of 156.7 ± 0.4°F (69.3 ± 0.2°C), the aniline need not be distilled.

Precautions for Handling Aniline

(1) At all times, avoid the inhalation of aniline vapors and handle the bottles containing the liquid with extreme care.

(2) If aniline should come into contact with the skin, wash the affected area *immediately* with soap and water. If aniline is

spilled on clothing, remove the article without delay and have it dry cleaned before reuse.

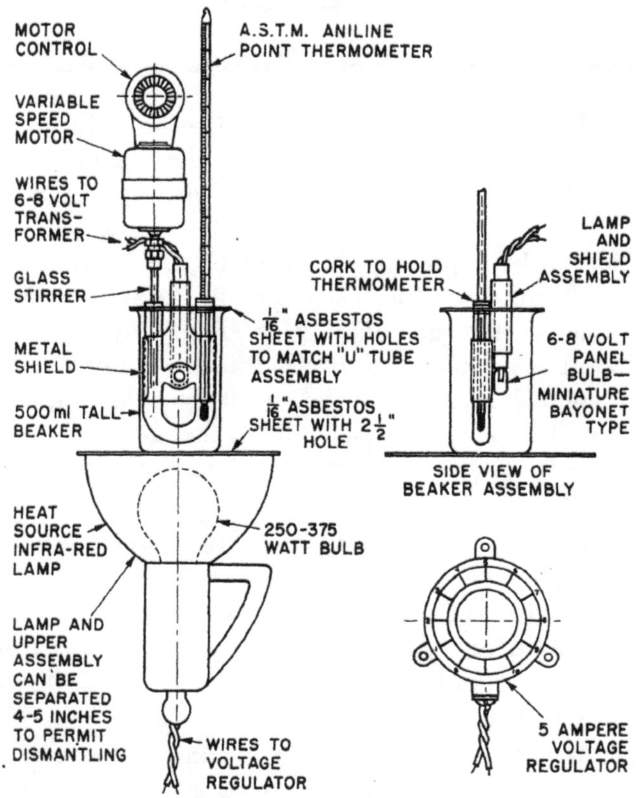

FIGURE 1. Aniline-Point Apparatus with U Tube

(3) If aniline is spilled on the floor or bench top, put on rubber gloves and mop up the spill immediately. Dispose of any material (cloth, sawdust, etc.) used in this operation in some manner other than simply throwing into a trash can. Avoid breathing aniline vapors while mopping up the spills.

(4) All operators handling aniline must be examined by the medical department before such work is started and must be re-examined at prescribed intervals.

Distillation of Aniline

Before assembling the purifying apparatus, put 100 ml of some liquid into the distillation flask and make a permanent mark on the side indicating the liquid level. Clean the flask and dry it in an oven. Put a few pellets of KOH into a bottle of aniline and reserve this particular bottle for this test. When the flask is dry, assemble the apparatus as shown in Figure 2. Insert the filling tube into the flask, put the bottle of aniline under the open end, turn on the aspirator, put a finger over the open end of the T-joint (shown on the drawing above the words TO ATMOSPHERE), and allow 100 ml of aniline to be drawn into the flask. Remove the tube, stopper the flask, and rinse the tube immediately. If the tube is to be used again on the same day, rinse with benzol; otherwise rinse with water. Apply heat so that the aniline will distill slowly. Discard the first 10-ml portion and continue heating until 78 ml have collected in the receiver.

As mentioned before, the aniline need not be distilled each day, but the distillate should not be stored in the receiver for more than 5 days.

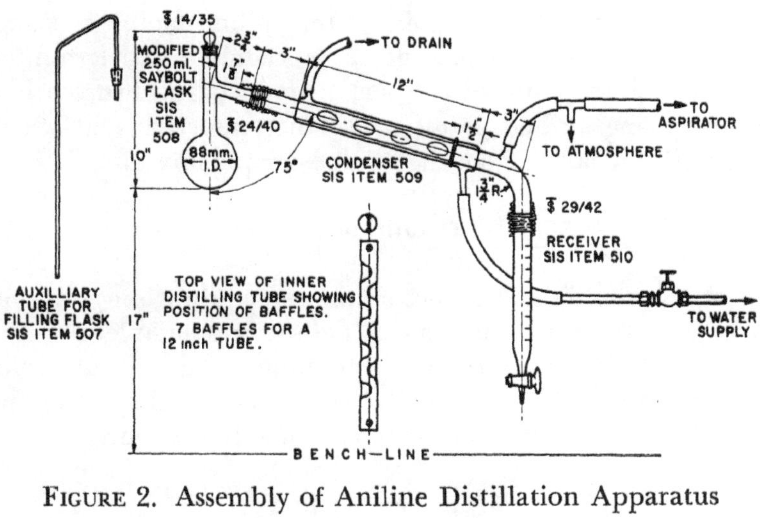

FIGURE 2. Assembly of Aniline Distillation Apparatus

Aniline Point

Put the required amounts of reagents into the U-tube. Put the proper thermometer into the right arm of the U, assemble the apparatus as shown in Figure 1, and put the stirrer into the left arm. The stirrer should turn in such direction that the liquid in the left arm will be lifted and forced through the connecting tube. Raise or lower the U-tube as necessary so that the cell will be in front of the small light bulb on the back of the holder. From here on, conduct the test as usual.

As previously stated, repeated experiments have shown that the results obtained with the apparatus described are comparable to those obtained with the thin-film assembly. Since the U-tube is mentioned in ASTM D–611 as an acceptable alternate, the existing items need not be replaced.

ASTM DESIGNATION: D 611–55 T

Scope

1. (*a*) This method describes the test procedures for determining the aniline point of petroleum products and hydrocarbon solvents, provided the aniline point is below the bubble point and above the solidification point of the aniline–sample mixture.

(*b*) This method also provides a procedure for determining the mixed aniline point of petroleum products and hydrocarbon solvents having aniline points below the temperature at which aniline will crystallize from the aniline–sample mixture.

Definitions

2. (*a*) *Aniline Point* is the minimum equilibrium solution temperature for equal volumes of aniline and sample.

(*b*) *Mixed Aniline Point* is the minimum equilibrium solution temperature of a mixture of 2 volumes of aniline, 1 volume of sample, and 1 volume of *n*-heptane of specified purity.

Outline of Method

3. Specified volumes of aniline and sample, or aniline and sample plus diluent, are placed in a tube and mixed mechanically. The mixture is heated at a controlled rate until the two phases become miscible. The mixture is then cooled at a controlled rate and the temperature at which the two phases separate is recorded as the aniline point or mixed aniline point.

Apparatus

4. (*a*) *Test Tube or Thin-Film Apparatus,* made of heat-resistant glass and conforming to the dimensions prescribed in the Appendix.[1]

(*b*) *Heating and Cooling Bath*—A suitable air bath, a non-aqueous, nonvolatile, transparent liquid bath, or an infrared lamp (250 to 375 w), provided with means for controlled heating is required.[2]

(*c*) *ASTM Aniline Point Thermometers,* having ranges of −36.5° to +107.5°F (−38° to +42°C), 77° to 221°F (25° to 105°C), and 194° to 338°F (90° to 170°C) and conforming to the requirements for thermometers 33°F (33°C), 34°F (34°C), and 35°F (35°C).

(*d*) Pipets, with capacities of 10 ± 0.04 ml and 5 ± 0.02 ml. Provide a rubber suction bulb for use with one of the 10-ml pipets when measuring aniline.

(*e*) *Balance*—A laboratory balance sensitive to 0.01 g, suitable for weighing the tube and sample when the sample cannot be pipetted conveniently.

Reagents

5. (*a*) *Aniline*[3]—Dry chemically pure aniline over potassium hydroxide pellets, decant, and distill fresh on the day of use,[4]

---

[1] Acceptable alternate apparatus may be used provided it has been shown to give results of the same accuracy and precision as those described in the Appendix. The U-tube equipment and the apparatus described in The Institute of Petroleum Method IP 2/53 are adequate for most samples covered by the scope of this method.

[2] Water should not be used as either a heating or cooling medium since aniline is hygroscopic and moist aniline will give erroneous test results. For example, the aniline point of the *n*-heptane reagent as measured with dry aniline is increased from that specified by approximately 0.9°F (0.5°C) if the aniline reagent contains 0.1 per cent water.

[3] Aniline should not be pipetted directly by mouth because of its extreme toxicity. Aniline is also toxic by absorption through the skin, even in *very small quantities,* and should be handled with great caution.

[4] As an alternative to distilling the aniline on the day of use, it may be distilled as described in paragraph (*a*), collecting the distillate in ampoules, sealing the ampoules under vacuum or dry nitrogen, and storing in a cool, dark place for future use. In either case, rigid precaution must be taken to avoid contamination from atmospheric moisture (see footnote 2). It is believed that under these conditions, the aniline will remain unchanged for a period exceeding 6 months.

discarding the first and last 10 per cent. Aniline thus prepared shall give an aniline point of 156.7° ± 0.4°F (69.3° ± 0.2°C) as determined from the average of two independent tests having a difference of not over 0.2°F (0.1°C) when tested with *n*-heptane according to Section 7.

(*b*) *Anhydrous Sodium Sulfate.*

(*c*) *n-Heptane,* conforming to the requirements listed in Table 1.

TABLE 1. Requirements for *n*-Heptane

Octane Number, ASTM Motor	−0.2 to +0.2
Density at 20°C, grams per milliliter	0.68375 ± 0.00015
Refractive Index, n_D^{20C}	1.38775 ± 0.00015
Freezing Point, °C	−90.72 minimum
Distillation: [5]	
Temperature of 50 per cent recovery, °C	98.43 ± 0.05
Temperature increase from 20 to 80 per cent recovery, °C	0.20 maximum

Sample

6. Dry the sample by shaking vigorously for 3 to 5 minutes with about 10 per cent by volume of a suitable drying agent, such as anhydrous calcium sulfate or anhydrous sodium sulfate. Reduce the viscosity of viscous samples by warming to a temperature below that which would cause the loss of light ends or the dehydration of the drying agent. Remove any suspended drying agent by centrifugation or filtration. When suspended water is visibly present and when the sample material is known to dissolve less than 0.03 per cent of water by weight, centrifugation for the removal of suspended water is an acceptable procedure.

[5] Determined by distilling 100 ml of sample through a 10-in. jacketed column. A platinum resistance thermometer is inserted so as to measure the condensation temperature of the vapor [see *Research Paper No. 482, Journal of Research,* Nat. Bureau Standards, **9,** No. 4, 459–461 (1932)]. Current through the jacket heating coil is adjusted so that the reflux from the thermometer is 5 to 6 drops per minute. The still-pot heater is adjusted to cause distillation to proceed at the rate of 5 ml per minute. The temperature is corrected to an atmospheric pressure of 760 mm of mercury by means of the equations of E. R. Smith and H. Matheson [see *Research Paper RP1097, Journal of Research,* Nat. Bureau Standards, **20,** 641 (1938)].

Procedure

7. Two methods, to be used as applicable, are covered as follows:

Method 1, described in detail in the Appendix, is applicable to clear, light-colored samples or to samples not darker than No. 6 ASTM color.

Method 2, described in detail in the Appendix, is applicable to light-colored samples, to moderately dark samples, and to very dark samples.[1]

Procedure for Mixed Aniline Point

8. This procedure is applicable to samples having aniline points below the temperature at which aniline crystallizes from the mixture. Pipet 10 ml of aniline (**Caution**—See footnote 3), 5 ml of sample, and 5 ml of *n*-heptane into a clean, dry apparatus. Determine the aniline point of the mixture by method 1 or 2 as described in the Appendix.

Report

9. (*a*) If the range of three successive observations of the aniline-point temperature is not greater than 0.2°F (0.1°C) for light-colored samples or 0.4°F (0.2°C) for dark samples, report the average temperature of these observations, corrected for thermometer calibration errors, to the nearest 0.1°F (0.05°C) as the aniline point.

(*b*) If such a range is not obtained after five observations, repeat the test using fresh quantities of aniline and sample in a clean, dry apparatus; and if consecutive temperature observations show a progressive change, or, if the range of observations is greater than the repeatability given in Section 10 (*a*), report the method as being inapplicable.

Precision

10. The following data should be used for judging the acceptability of results (95 per cent probability):

(*a*) *Repeatability*—Duplicate results, that is, two average temperatures obtained by the same operator in a series of observations as described under Section 9, should be considered suspect if they differ by more than the amounts listed in Table 2.

TABLE 2

	Repeatability
Aniline Point of:	
Clear, light-colored samples	0.3°F (0.16°C)
Moderately dark to very dark samples	0.6°F (0.3°C)
Mixed Aniline Point of:	
Clear, light-colored samples	0.3°F (0.16°C)
Moderately dark to very dark samples	0.6°F (0.3°C)

(*b*) *Reproducibility*—When a result is submitted by each of two laboratories, these two results should not be considered suspect unless they differ by more than amounts listed in Table 3.

TABLE 3

	Reproducibility
Aniline Point of:	
Clear, light-colored samples	0.9°F (0.5°C)
Moderately dark to very dark samples	1.8°F (1.0°C)
Mixed Aniline Point of:	
Clear, light-colored samples	1.3°F (0.7°C)
Moderately dark to very dark samples	1.8°F (1.0°C)

APPENDIX

Method 1

A1. *Apparatus*—The apparatus shown in Figure 3 shall consist of the following:

(*a*) *Test-Tube*—A test-tube approximately 25 mm in diameter and 150 mm in length, made of heat-resistant glass.

(*b*) *Jacket*—A jacket approximately 37 to 42 mm in diameter and 175 mm in length, made of heat-resistant glass.

(*c*) *Stirrer*—A manually operated metal stirrer of No. 14 B & S gage soft iron wire as shown in Figure 3. A concentric ring shall be at the bottom, having a diameter of approximately 19 mm.

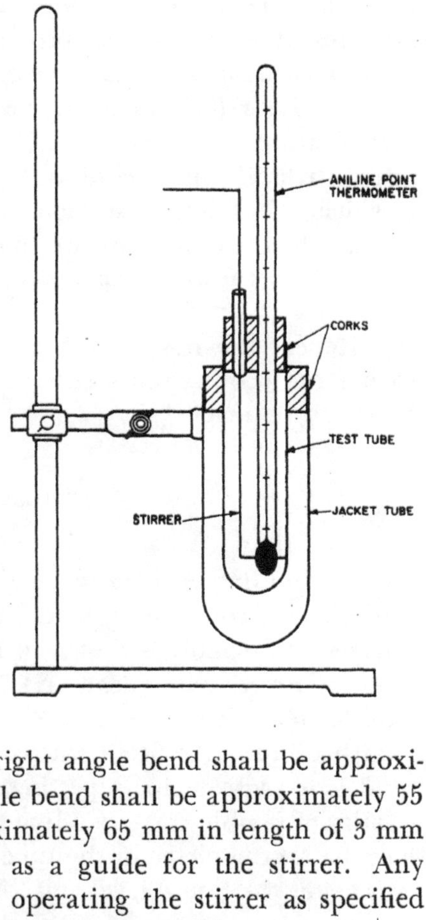

FIGURE 3.
Aniline-Point
Apparatus

The length of the stirrer to a right angle bend shall be approximately 200 mm. The right angle bend shall be approximately 55 mm long. A glass sleeve approximately 65 mm in length of 3 mm inside diameter shall be used as a guide for the stirrer. Any suitable mechanical device for operating the stirrer as specified is an approved alternate for the manual operation.

A2. (*a*) *Procedure*—Clean and dry the apparatus. Pipet 10 ml of aniline (**Caution**—see footnote 3) and 10 ml of the dry sample into the air-jacketed tube fitted with stirrer and thermometer. If the liquid is too viscous for pipetting, weigh to the nearest 0.01 g a quantity of the sample corresponding to 10 ml at room temperature. Center the thermometer in the test tube so that the immersion mark is at the liquid level, making sure that the bulb does not touch the side of the tube.

(*b*) Stir the sample rapidly, using a 2-in. stroke, avoiding the inclusion of air bubbles and, if necessary, heating at a rate of 2° to 5°F (1° to 3°C) per minute until complete miscibility is obtained, by applying heat directly to the jacket tube. If the aniline sample mixture is completely miscible at room temperature, substitute a nonaqueous cooling bath for the heating source. Continue stirring and allow the mixture to cool slowly at a rate of 1.0° to 1.8°F (0.5° to 1.0°C) per minute. Continue cooling to a temperature of 2.0° to 3.5°F (1° to 2°C) below the first appearance of turbidity and record as the aniline point the temperature at which the mixture suddenly becomes cloudy throughout.[6] This will be the minimum equilibrium solution temperature and not the temperature of separation of minor portions of the samples.

(*c*) Repeat the observation of aniline point temperature by heating and cooling repeatedly until a report as directed in Section 9 can be made.

Method 2

A3. *Thin-Film Apparatus* made of heat-resistant glass and stainless steel, conforming to the dimensions given in Figure 4. A suggested assembly is shown in Figure 5.

A4. (*a*) *Procedure*—Clean and dry the apparatus. Pipet 10 ml of aniline (**Caution**—See footnote 3) and 10 ml of the sample into the air-jacketed tube fitted with stirrer and thermometer. If the liquid is too viscous for pipetting, weigh to the nearest 0.01 g a quantity of sample corresponding to 10 ml at room temperature. Place the thermometer in the tube so that the contraction chamber is below the liquid level and the mercury bulb does not touch the side of the tube.

(*b*) Adjust the speed of the pump to produce a continuous stream of the oil–aniline mixture in the form of a thin film flowing over the light well. With extremely dark oils, operate the pump slowly and lower it so that the delivery tube nearly touches the top of the light well, thus obtaining a film that is continuous

---

[6] The true analine point is characterized by a turbidity which increases sharply as the temperature is lowered.

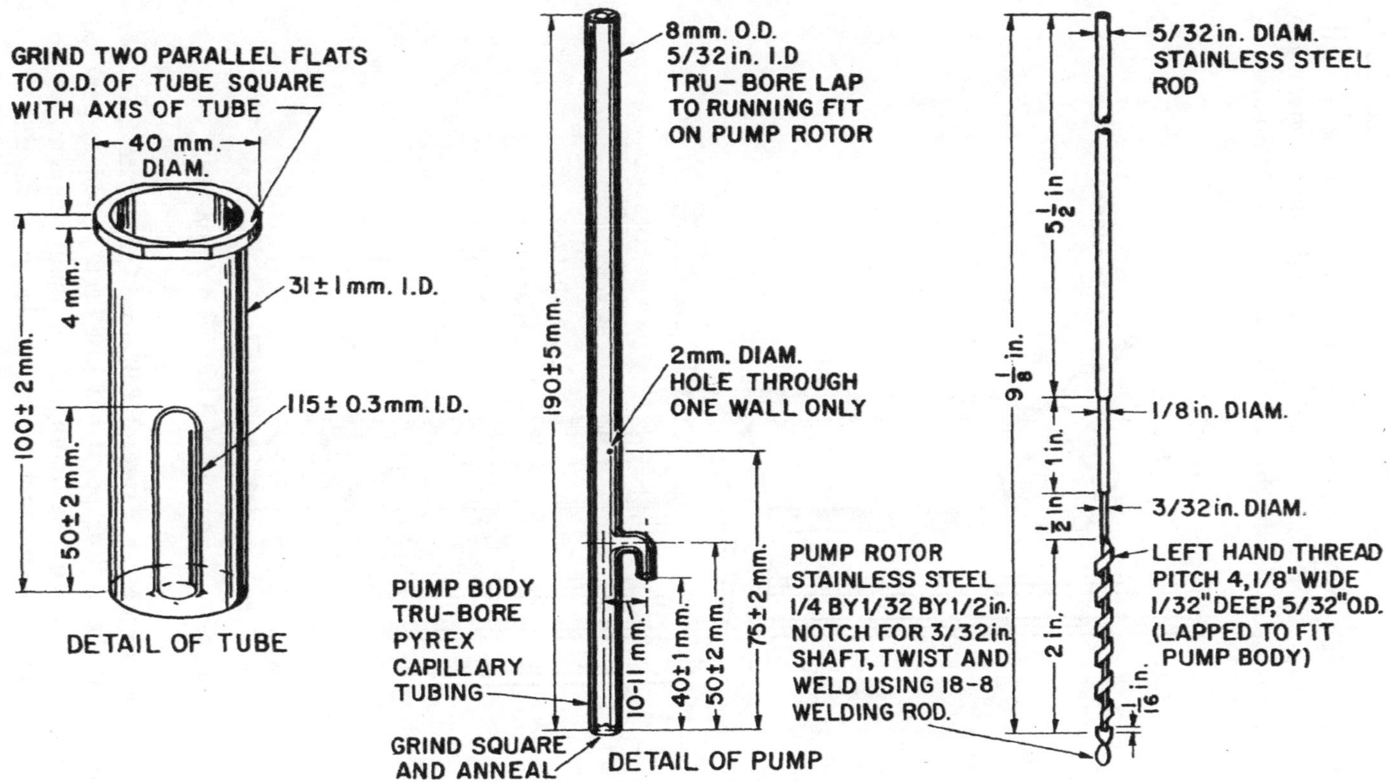

FIGURE 4. Details of Aniline Point Thin-Film Apparatus

and thin enough to permit observation of the aniline point. Adjust the voltage on the lamp until just enough light is given for the filament to be visible through the film. Raise the tempera-

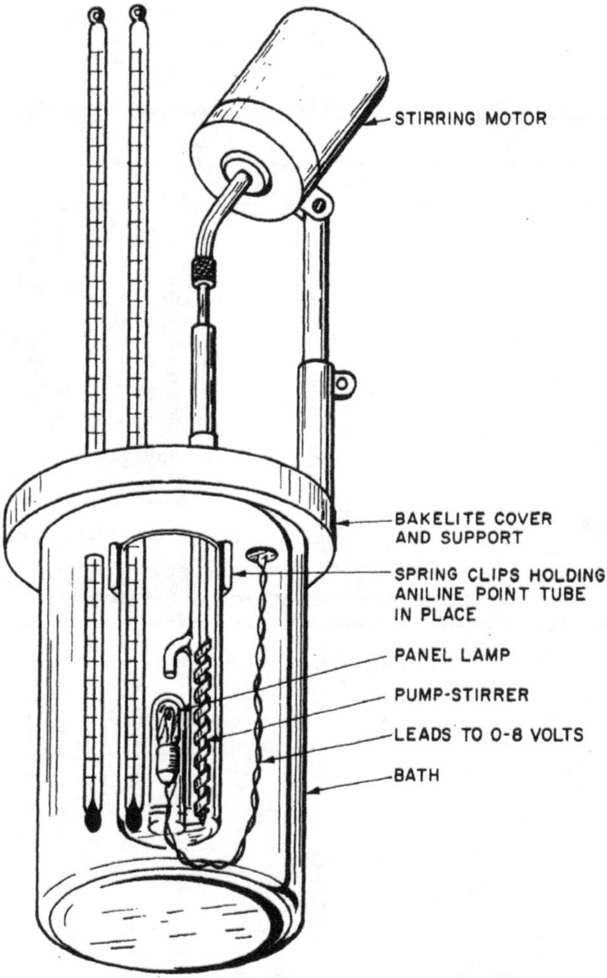

FIGURE 5. Assembly of Thin-Film Apparatus

ture of the mixture at a rate of 2.0° to 3.5°F (1° to 2°C) per minute until the aniline point has just been passed as denoted by a definite, sudden brightening of the lamp filament, and by the disappearance of the more or less opalescent condition of the film.[7] Discontinue heating and adjust the lamp voltage so that the filament appears clear and distinct but not uncomfortably bright to the eye. Adjust the temperature of the bath so that the sample–aniline mixture cools at a rate of 1.0° to 1.8°F (0.5° to 1.0°C) per minute and note the appearance of the film and light filament. Record as the aniline point the temperature at which a second phase appears as evidenced by the reappearance of the opalescent condition of the film (usually causing a halo to appear around the lamp filament) or by a sudden dimming of the lamp filament, or both.

(*c*) Repeat the observation of aniline-point temperature by heating and cooling repeatedly until a report as directed in Section 9 can be made.

---

[7] For those making the test for the first time, the following procedure may be helpful: Make preliminary operational adjustments and tests, using a colorless sample–aniline mixture and observing the changes taking place in the body of the liquid and film. Make rough tests with dark oils to become familiar with the appearance of the film and light source as the mixture passes from the clear state above the aniline point to the translucent state below. If the sample is such that there is difficulty in observing the exact point of the phase change, make experiments with the sample, using various intensities of light and paying particular attention to the appearance of the light in the immediate vicinity of the lamp filament.

WIJS' IODINE NUMBER

This method provides a means of approximating the amount of unsaturation in an oil. The sample under test is treated with a measured amount of a standard solution of iodine monochloride, and the excess determined by titration with standard thiosulfate. During the treatment, any unsaturation in the sample under test will be filled by the halogen. Thus, the halogen consumed, calculated as iodine, is a measure of the unsaturation of the sample.

The details of the test are extremely important, since there may be substitution as well as addition. Further, the details of the test depend on the type of sample under examination. This method will, therefore, be divided into two sections. One is intended for the examination of fats and fatty oils and is based on similar procedures of the American Oil Chemists' Society. The other is for the examination of petroleum oils, and is based on experimental work done in the Enjay laboratories.

Definition

The iodine value or iodine number is the number of grams of halogen, calculated as iodine, which is absorbed by 100 g of the sample under test conditions.

Reagents

(1) *Iodine Solution*—Three different methods are listed for preparing the solution. The first is reprinted from the American Oil Chemists' Society procedure, and must be followed for referee or precision testing. Either of the others may be used for routine testing or plant-control work. Whatever the method of preparation, the following precautions must be rigidly observed.

(*a*) At all times, during preparation and subsequent storage, the reagents must be protected from moisture.

(*b*) The glacial acetic acid must be of American Chemical Society reagent grade.

(*c*) The final solution should have a *slight* excess of iodine, be stored in an amber bottle, and contain the equivalent of 0.026 ± 0.001 g of iodine per milliliter.

Method A

Dissolve 13 g of iodine (ACS grade) in 1 liter of glacial acetic acid, heating as necessary to assist solution. After cooling, bubble in dry chlorine gas until the original concentration of iodine has been almost, but not quite, doubled. Store the final reagent in an amber bottle and protect from atmospheric moisture.

A distinct change of color will be noted as the end point is neared. It is convenient to reserve a small amount of the original solution, introduce chlorine into the remainder until the end point has been passed, and then add the retained portion as necessary. The chlorine may be obtained either from a cylinder or by the action of hydrochloric acid on potassium permanganate or a mixture of permanganate and manganese dioxide. In any case, the chlorine must be dried by bubbling through concentrated sulfuric acid. The strength of the solution is determined as described under PROCEDURE, (3) to (6).

Method B

The reagent may be purchased, ready to be used, from a chemical supply house (e.g., Fisher Scientific Company).

Method C

Dissolve 16.6 g of pure iodine monochloride (available from the Eastman Kodak Company) in 1 liter of glacial acetic acid.

(2) *Potassium Dichromate*—A standard sample with a certificate of analysis is available from the National Bureau of Standards. It is strongly recommended that this sample is used as a primary standard and a 0.1 N solution prepared, following the directions on the label.

For routine work, a 0.1 N solution may be prepared by dissolving 4.9035 g of ACS-grade reagent, previously ground to a fine powder and dried, in distilled water and diluting to 1 liter.

(3) *Potassium Iodide*—Dissolve 150 g of the reagent in distilled water and dilute to 1 liter.

(4) *Starch Solution*—Make a homogeneous paste of 10 g of the soluble starch in cold distilled water, add 1 liter of boiling distilled water, and stir rapidly. After cooling, add 1.25 g of salicylic acid. When not in use, store in the solution in an ice box.

If preferred, the starch may be freshly prepared for each test, following the procedure outlined under (4), but omitting the salicyclic acid. Use only 1 g of the starch, dissolving it in 100 ml of hot water.

(5) *Sodium Thiosulfate*—Dissolve 25 g of the chemically pure pentahydrate (or 15.9 g of the anhydrous salt) in 1 liter of freshly boiled and cooled distilled water and allow the solution to stand for 1 week in a tightly stoppered container.

To standardize, measure into a clean iodine flask or bottle about 10 ml of the potassium iodide and exactly 25 ml of the 0.1 N dichromate solutions. Add 5 ml of concentrated hydrochloric acid and, without delay, titrate the mixture with the thiosulfate, adding 5 ml of the starch solution as the color fades and continuing to the complete disappearance of the blue color. Calculate the normality of the thiosulfate solution by the equation:

$$\text{Normality of thiosulfate} = \frac{(\text{Milliliters of dichromate}) \times (\text{Normality of dichromate})}{\text{Milliliters of thiosulfate solution}}$$

Apparatus

(1) *Flask*—The recommended container is a 500-ml iodine-number flask, which may be obtained from any regular supply house. A permissible alternate is a 500-ml glass-stoppered tincture bottle.

Procedure for Fats and Fatty Oils

(1) Melt the sample if it is not a liquid, heating to a temperature not higher than about 15°C above the melting point and stirring continuously. Filter the sample through paper to remove any impurities.

(2) Weigh accurately the proper amount of the sample into a 500-ml flask or bottle containing 20 ml of carbon tetrachloride.

The proper amount of sample depends on the expected iodine number and should be such that the excess of iodine will be between 100 and 150 per cent of the amount absorbed. Table 1 may be used as a guide.

TABLE 1. Correct Amount of Sample

Expected Iodine Number	Maximum in grams	Minimum in grams	Accuracy of Weighing
Below 3		10+	±0.0010
3	10.500	8.500	0.0050
5	6.400	5.100	0.0005
10	3.100	2.500	0.0002
20	1.500	0.850	0.0002
40	0.790	0.630	0.0002
60	0.530	0.420	0.0002
80	0.400	0.310	0.0001
100	0.320	0.250	0.0001
120	0.260	0.210	0.0001
140	0.227	0.181	0.0001
160	0.198	0.159	0.0001
180	0.176	0.141	0.0001
200	0.159	0.127	0.0001

Table 1 is applicable to normal, nondrying fats and fatty oils. For drying oils, the excess should be between 115 and 135 per cent of the amount absorbed. When the sample is a drying oil the maxima should be decreased and the minima increased by about 10 per cent.

(3) Pipette 25 ml of the iodine solution into the flask containing the sample and swirl to insure complete mixture.

(4) Store the flasks in a dark place for 30 minutes at 25° ± 5°C. (When the expected number is above 150, store for 1 hour.) Include in the set at least two blanks containing the same amounts of carbon tetrachloride and iodine solution, but no sample.

(5) Remove the flasks from storage and add to each 20 ml of the potassium iodide solution, followed by 100 ml of distilled water.

(6) Titrate each with the 0.1 N thiosulfate solution, adding it slowly and with constant, vigorous agitation. Continue until the

yellow color of iodine has almost disappeared. Add 1 to 2 ml of the starch solution and continue to the complete disappearance of the blue color. Note and record the volume of thiosulfate solution.

The swirling must be continuous and vigorous during the titration. While not mandatory, the use of some type of mechanical stirrer is suggested.

Calculations

Calculate the iodine number by the equation:

$$I = \frac{(B - S) \times N \times 12.69}{W}$$

where

- B is the volume of the thiosulfate solution required for the blank, in milliliters.
- S is the volume of thiosulfate required for the sample.
- N is the exact normality of the iodine solution.
- W is the weight of the sample, in grams.
- I is the iodine number, or value, of the sample under test.

Procedure for Petroleum Oils

The procedure is generally similar to that used for fats, with the following modifications:

(*a*) The weight of sample should be such that the amount of iodine consumed will be not less than 10 nor more than 20 per cent of the amount added originally. If the nature of the sample is unknown, a trial test may be necessary.

(*b*) Some products dissolve very slowly in the carbon tetrachloride. If necessary, the sample should be allowed to stand, with occasional shaking, until solution is complete before proceeding.

(*c*) The mixture of sample, solvent, and iodine solution should be allowed to stand for 30 ± 1 minute before adding the potassium iodide solution. Since this time is critical, the samples should be set up at 5- to 10-minute intervals, to allow each to stand for the exact time before titrating with the thiosulfate.

During the early stages of the titration with thiosulfate, the flask should be swirled, rather than inverted and shaken, to avoid loss of the iodine solution from around the stopper by spurting. After the starch has been added, this consideration becomes less important. If a mechanical stirrer is not used, the flask may then be stoppered and shaken vigorously; any spurting drops will be caught in the neck of the flask and may be washed back with a few milliliters of distilled water.

Petroleum oils do not have a constant iodine number; the observed value will vary sharply with the excess of iodine and the time of standing. The test conditions must be followed exactly if reproducible results are to be obtained.

When the sample under test is a tung oil, the weight of sample should be 0.15 ± 0.05 g, and the prepared flask should be allowed to stand for 1 hour at a 20° to 25°C.

BROMINE NUMBER (LEWIS AND BRADSTREET METHOD)

This method is used for the determination of unsaturates in petroleum mixtures and is applicable to straight- and branched-chain olefines of low molecular weight. With highly branched polymers, or in the presence of certain sulfur compounds, the results are erratic. Conjugated diolefines show a bromine number corresponding to one half of the theoretical value.

Apparatus

(1) *Reaction Flask*—Either a 200-ml glass-stoppered iodine flask or a 500-ml glass-stoppered tincture bottle may be used.

(2) *Weighing Pipette*—The preferred type is the Lunge weighing pipette, equipped with two glass stopcocks. If not available, a small weighing bottle, fitted with a rubber nipple, may be substituted.

Reagents

(1) *Potassium Bromide Solution*—A 10 per cent by volume solution of sulfuric acid in distilled water is saturated with chemically pure potassium bromide.

(2) *Sodium Thiosulfate Solution*—Dissolve 24.8 g of the chemically pure pentahydrate in 1 liter of recently boiled and cooled distilled water. Allow the resulting solution to stand for 1 week and then standardize, using an accepted analytical method.

(3) *Bromate–Bromide Solution*—Dissolve 50 g of KBr and 14 g of $KBrO_3$ in sufficient distilled water to make 1 liter. Standardize against the thiosulfate in the manner described under PROCEDURE, omitting the addition of the sample.

(4) *Saturated Potassium Iodide Solution*—Use the chemically pure salt with a very low iodate content. If a light-yellow color develops, destroy it by the addition of 1 or 2 drops (avoiding excess) of the thiosulfate solution.

Procedure

(1) Place 20 ml of the KBr solution and 15 ml of *n*-heptane or other suitable solvent into a reaction flask. If mercaptans and/or disulfides are present in the sample, add also 1 ml of a 10 per cent aqueous solution of zinc sulfate.

(2) Add 0.7 to 1.0 g of sample from the weighing pipette and stopper the flask immediately. Titrate the contents of the flask with the bromate–bromide solution, swirling constantly, to a faint yellow color. Add exactly 1 ml in excess of $KBr–KBrO_3$ and shake the flask vigorously for 2 minutes. Note and record the exact volume added.

(3) Add 5 ml of satuated KI solution and titrate the liberated iodine with the standard thiosulfate solution, using potato starch as the indicator. Record also the volume of thiosulfate.

Calculations

The bromine number, expressed in centrigrams of bromine per gram of sample, may be calculated as follows:

$$A = \frac{(BC - DE) \times 7.99}{W}$$

where

A is the bromine number.
B is the normality of bromate–bromide solution.
C is the volume of bromate–bromide solution.
D is the normality of thiosulfate solution.
E is the volume of thiosulfate solution.
W is the weight of sample.

L.I.C. No. 180.13

DETERMINATION OF THE BROMINE INDEX OF XYLENE

The usual methods for determining bromine number are not satisfactory when applied to products such as xylene, since the experimental error is generally at least as large as the bromine number of the sample. The ASTM has investigated this subject and developed a special method which has been published for information only.

The new method is generally similar to the electrometric procedure described in ASTM D 1159, the major difference being that the results are reported in terms of the bromine index, rather than the bromine number. The numerical value of the index is one thousand times that of the number. Since there is such a large difference in these terms, the report should state clearly that the bromine index is being reported, not the number.

METHOD OF TEST FOR BROMINE INDEX OF AROMATIC HYDROCARBONS BY POTENTIOMETRIC TITRATION [1]

Scope

1. This method covers a procedure for determination of the bromine index of aromatic hydrocarbons by potentiometric titration.

Outline of Method

2. The sample is dissolved in a special solvent and titrated with 0.02 *N* bromide–bromate solution at room temperature. Shielding from light is not necessary. End-point detection is by

[1] This proposed method is under the jurisdiction of the ASTM Committee D-16 on Industrial Aromatic Hydrocarbons and Related Materials.

a potentiometric method using the combination of glass *versus* platinum electrodes.

Apparatus

3. (*a*) *Titrimeter or pH Meter* equipped with glass and platinum electrodes. The pair of electrodes shall be mounted to extend well below the liquid level.

(*b*) *Buret,* 25 ml.

(*c*) *Mechanical Stirrer* that will furnish a rapid stirring action, but not such vigorous sitrring that air bubbles will be drawn down to the electrodes of the titrimeter or *p*H meter.

(*d*) *Beaker.*

Reagents

4. (*a*) *Titration Solvent* conforming to Section 5 (*a*) of ASTM Designation D 1159.

(*b*) *Standard Potassium Bromide–Bromate Solution (0.02 N)*—Prepare and standardize as directed in Section 5 (*b*) of method D 1159, but make suitable adjustments for a 0.02 *N* solution instead of 0.5 *N*.

Procedure

5. (*a*) Place 100 ml of titration solvent in the titration beaker and either weigh or pipet the appropriate size sample into the solvent. Use Table 1 as a guide in choice of sample size.[2]

TABLE 1

Bromine Index	Grams of Sample
0 to 10	25 to 30
10 to 50	15 to 25
50 to 100	10 to 25
100	10

[2] The size of sample is limited by the tolerance of the solvent–hydrocarbon sample mixture for water. If there is too much hydrocarbon or too large a titration with aqueous bromide–bromate solution, the titration mixture may eventually separate into two phases. If the mixture becomes cloudy or separates into two phases, discard the analysis and repeat with a smaller sample.

(*b*) Set the beaker in place on the titration stand, insert the glass and platinum electrodes, start the stirrer, and commence the titration with 0.02 N bromide–bromate solution. Conduct the titration in a manner similar to the usual potentiometric titrations, taking readings of emf *versus* milliliters of titrant. Take care that the rate of addition of bromide–bromate solution does not exceed 0.5 ml per minute. As the end point is approached, add the reagent in 0.2 ± 0.05-ml increments and take the corresponding emf reading. Consider the end point to be reached when a 0.2-ml increment of reagent causes a deflection of at least 0.075, based on the maximum deflection after the addition of the reagent.

(*c*) Run a titration blank on the solvent, following the same procedure, but omitting the sample.

Calculation

6. The bromine index shall be calculated as follows:

$$\text{Bromine index} = \frac{(\text{A} - \text{B}) \times N \times 7990}{\text{W}}$$

where

A is the milliliters of bromide–bromate solution used for titration of the sample.
B is the milliliters of bromide–bromate solution used for titration of the blank.
N is the normality of the bromide–bromate solution.
W is the grams of sample used.

L.I.C. No. 180.52

DETERMINATION OF THE AVAILABLE CYCLOPENTADIENE AND METHYLCYCLOPENTADIENES IN DIMERS

This method is based on a similar one appearing in the literature.[1] The results obtainable with this procedure are strictly empirical, and it is essential that the details outlined are followed exactly.

The scope has not been completely determined, since it is known that certain compounds such as indenes and higher homologues of cyclopentadiene will materially affect the determination of cyclopentadiene. However, the observed results are accurate to about 1 per cent, if the sample under test contains at least 85 per cent of the dimers and codimers of cyclopentadiene and the isomeric methylcyclopentadienes.

A portion of the sample under test is mixed with isooctane and the solution passed through the cracking apparatus shown in Figure 1. By adjusting the heating conditions so that the vapor leaving the cracker is at 680° ± 10°F, both dimers will be converted to the original cycle compounds. The vapors leaving the cracker are condensed to liquids, diluted to volume with more isooctane, and the final solution examined in an ultraviolet spectrophotometer. By comparing the absorption of the solution with that of calibration standards treated in a similar manner, the amounts of cyclopentadiene and methylcyclopentadiene in the sample can be determined.

Apparatus

(1) *Depolymerizer* as shown in Figure 1. Provision must be made for measuring the temperature of the vapors at the bottom by a thermocouple.

(2) *Spectrophotometer*—The Beckman Model DU, equipped

[1] *Anal. Chem.*, **20,** 510 (1948).

with 1-cm cells, is the recommended instrument. Any other may be used, providing that readings of absorption can be made at settings of 240 and 258 millimicrons.

(3) *Volumetric Flasks* of 50-ml and 100-ml capacity.

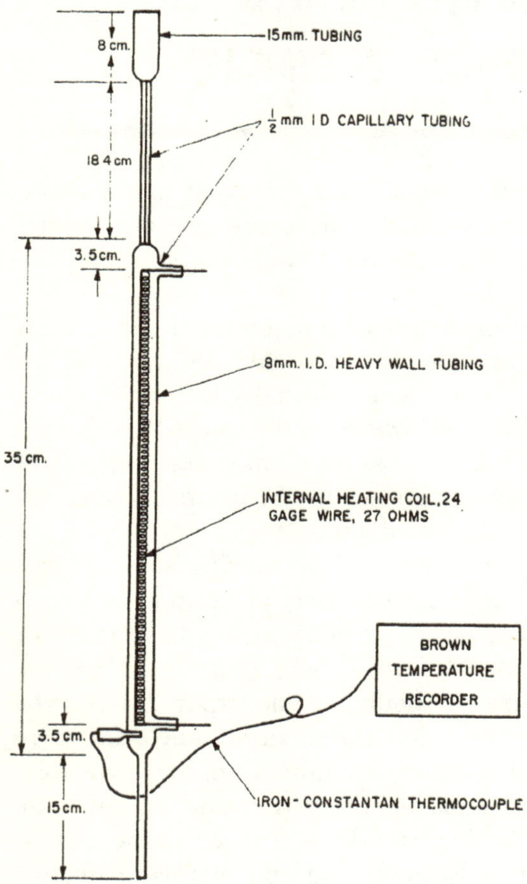

FIGURE 1. Dicyclopentadiene Depolymerization Apparatus

Reagents

(1) *Isooctane* of spectroscopic quality, available from the Phillips Petroleum Company. Before use, examine a portion in the spectrophotometer at settings of both 240 and 258 millimicrons (mμ). If it shows any absorption at either setting, pass it through activated silica gel (20 to 200-mesh) until no absorption is shown.

(2) *Dimer of Cyclopentadiene* of satisfactory purity is available from Matheson, Coleman and Bell, East Rutherford, N. J., under their catalog number 5757.

(3) *Dimer of Methylcyclopentadienes*—No suitable product, either monomer or dimer, is available commercially. To prepare a standard, distill in a vacuum apparatus a petroleum stream containing the monomer, separating and collecting the cut which boils between 2° and 15°C at 50 mm pressure. Store this cut in a tightly sealed container made of metal or heavy-walled glass and heat for 24 hours at 90°C. Again distill under vacuum, discarding the portion which boils below 110°C. Further purify the bottoms by two more vacuum distillations, in each case taking a heart cut which boils between 111° and 114°C at 50 mm pressure. The final product should be a water-white liquid showing the physical constants listed in Table 1.

TABLE 1

	Minimum	Maximum
Refractive Index n_D^{20}	1.4991	1.4998
Specific Gravity 20/20	0.9413	0.9416
Boiling Point at 50 mm pressure	110°C	114°C

Calibration

(1) Weigh accurately about 5 g of the pure dimer of cyclopentadiene into a tared 100-ml volumetric flask. Partly fill the flask with isooctane, shake until solution is complete, and fill to the mark.

(2) Apply current to the coil in the depolymerizer. Put a receiving flask under the outlet, mounting the flask in a bath filled with alcohol or other suitable solvent. By means of dry ice, cool the bath to a temperature not higher than 0°F. Pour about 5 ml of isooctane into the top of the apparatus, and adjust the heating rate so that the temperature of the vapors at the bottom will be 680° ± 10°F, adding more isooctane through the top as necessary.

(3) When the heating rate has been adjusted, remove the re-

ceiver and replace with a 100-ml volumetric flask containing about 30 ml of isooctane. Allow any isooctane remaining in the funnel to pass over the wire and collect the condensate. When the last of the isooctane has passed into the capillary of the funnel, pour 5 ml of the diluted sample into the funnel and allow this solution to pass over the hot wire. When the last of this solution has passed into the capillary, add another 5 ml of isooctane, washing down the sides of the funnel with the liquid.

(4) When the last of the final wash has entered the capillary, remove the receiver, stopper it, and allow to stand until the contents have reached room temperature. Fill the flask to the mark with isooctane and mix thoroughly. Pipette 5 ml of this solution into a 100-ml flask, fill to the mark, and mix thoroughly. Dilute again in the ratio of 5 to 50.

(5) Transfer a portion of the final solution to a 1-cm cell and scan in the general area of 235 to 260 mμ. Find by experiment the exact location of the peak heights at or near 240 or 258 mμ and calculate the absorptivity at each setting by the equation:

$$a = \frac{A}{CL}$$

where

- A is the net observed absorbancy (optical density) at each setting, as compared with a reference cell filled with isooctane.
- C is the concentration of the sample, in grams per liter of solution.
- L is the optical path, in centimeters (may be omitted if the same cells are used for calibration and analysis).
- a is the absorptivity (extinction coefficient) of the sample under test.

(6) Repeat the entire procedure using the pure dimer of methylcyclopentadienes and calculate the "a" values at both settings.

While the actual values may vary with the particular instrument in use, those listed in Table 2 were noted in the development of this method.

TABLE 2. "a" Values

Compound	240 mμ	258 mμ
Cyclopentadiene	46.5	17.1
Methylcyclopentadienes	38.1	28.2

Procedure

(1) Pipette 5 ml of the sample into each of two tared 100-ml volumetric flasks and reweigh. Note the weight of sample and fill both flasks to the mark with isooctane.

(2) Reserve one of the flasks as a reference standard and treat the solution in the other in the same manner as under CALIBRATION. The final dilution [see (4)] should be at the rate of 5 to 50 only if the sample under test contains at least 85 per cent of dimer. If the sample contains less than 85 per cent, this dilution ratio should be 10 to 50. Whatever the ratio, the reference sample should be diluted in the same manner.

(3) Determine the absorbancy of the cracked sample at the two critical settings, using the diluted solution from the second flask as a reference. Find the absorptivity at each setting by the same equation, and calculate the amount of dienes from:

Weight Per Cent Cyclopentadiene $= (4.27 \times a_{240}) - (5.77 \times a_{258})$

Weight Per Cent Methylcyclopentadiene $= (7.05 \times a_{258}) - (2.59 \times a_{240})$

The values 4.27, 5.77, etc., are valid only if the actual "a" values are 46.5 at 240 mμ, 17.1 at 258 mμ, and so on. If the "a" values differ markedly from those listed, the corresponding factors must be calculated and substituted in the equations. To do this, first determine a generic factor F by the equation:

$$F = \frac{1 \times 100}{a_1a_4 - a_2a_3} \quad \ldots \text{(a)}$$

where

a_1 is the actual absorptivity at 240 mμ of cyclopentadiene.

a_2 is the actual absorptivity at 240 mμ of methylcyclopentadiene.

a_3 is the actual absorptivity at 258 mμ of cyclopentadiene.
a_4 is the actual absorptivity at 258 mμ of methylcyclopentadiene.

Substituting the corresponding values into equation (a):

$$F = \frac{100}{(46.5 \times 28.2) - (38.1 \times 17.1)} \text{ or } 0.15136$$

Then calculate each of the other values as follows:

$$F_1 = F \times a_4 \text{ or } (0.15136 \times 28.2 = 4.27)$$
$$F_2 = F \times a_2 \quad (0.15136 \times 38.1 = 5.77)$$
$$F_3 = F \times a_1 \quad (0.15136 \times 46.5 = 7.05)$$
$$F_4 = F \times a_3 \quad (0.15136 \times 17.1 = 2.59)$$

Substitute these values in the proper equations and calculate the weight per cent of each component.

L.I.C. No. 180.70

DETERMINATION OF CYCLOPENTADIENE IN ISOPRENE

This procedure is based on unit 7.12.10 of the *Laboratory Manual* issued by the Office of Rubber Reserve.

The method is intended for the determination of cyclopentadiene present as the monomer in isoprene. It may be applied to samples containing up to 5 per cent cyclopentadiene and also for higher concentrations but with less accuracy.

Aldehydes and ketones interfere, but can be removed by water washing. Aromatics and tertiary butyl catechol, in high concentrations, also interfere but can be removed by distillation.

In an alkaline medium, cyclopentadiene reacts with both aldehydes and ketones to give colored compounds known as fulvenes. The color developed is dependent on the amount of cyclopentadiene present and the type of compound used to bring about the reaction. By selecting a specific reactor (in this case, benzaldehyde) and comparing the developed color with a calibration curve, the amount of cyclopentadiene in the sample can be determined.

Apparatus

Colorimeter—Any convenient type, grating or filter, may be used providing that a measurement of color can be made in the neighborhood of 450 mμ.

Reagents

(1) *Cyclopentadiene*—The reagent used for tracing the calibration curve can be prepared by cracking dicyclopentadiene at its boiling point in a simple still in the presence of an antioxidant such as *p*-tertiary butyl catechol and in a nitrogen atmosphere. Since the monomer will readily revert to the dimer, the calibration curve should be prepared within a few hours after the

distillation. This dimerization can be retarded by keeping the monomer cold.

(2) *Alcoholic Benzaldehyde Solution*—Distill USP benzaldehyde through a 1-foot Vigreaux column, collecting a cut with a boiling range of 179° to 180°C. Mix 20 ml of the distillate with enough ethanol (SDA–2B) to make 100 ml of solution. Protect the final solution from unnecessary exposure to the air.

(3) *Alcoholic Potassium Hydroxide*—Dissolve 3 g of the chemically pure reagent in 97 g of ethanol (SDA–2B). Filter the solution before use.

Calibration

(1) Weigh 1.00 ± 0.01 g of the pure cyclopentadiene into a 500-ml volumetric flask containing ethanol and dilute to the mark. Mix well and store in a cool place. The solution will contain 0.002 g of cyclopentadiene in 1 ml.

(2) Using a pipette or burette, measure 1.0, 2.0, 3.0 and 4.0 ml of the solution into separate Erlenmeyer flasks. Dilute each to 5.0 ml with ethanol, and put 5.0 ml of ethanol into another flask to serve as a blank.

(3) Add 5.0 ml of the alcoholic KOH and 5 ml of the 20 per cent benzaldehyde solution into each flask, including the blank. Swirl to mix the contents and allow to stand for 5 minutes.

(4) Fill the reference cell of the colorimeter with the solution representing the blank and another cell with one of the sample solutions. Put both cells into the colorimeter and determine the optical density (absorbancy) of the sample *versus* the blank at a setting of about 450 mμ.

The operating instructions for the colorimeter will depend on the type used. The operator should find by experiment the optimum testing conditions and use these settings for all future work. The exact setting of 450 mμ is only suggested, not mandatory.

(5) Repeat the procedure with each of the other standard samples, determining in each case the optical density as compared with the blank. Prepare a graph, plotting optical density against weight of cyclopentadiene in the test solution.

Procedure

(1) Weigh a glass-stoppered 50-ml graduate containing about 25 ml ethanol on an analytical balance to the nearest milligram. Remove the graduate, add the amount of sample shown in Table 1, and weigh again. Record the weight of the sample in the graduate.

TABLE 1

Percentage of Cyclopentadiene in Sample	Milliliters of Sample to Be Used
0.0 to 1.0	5.0
1.0 to 2.5	2.5
2.5 to 5.0	1.0

Since isoprene has a boiling point of 34°C, the sample should be thoroughly chilled before the test is started.

(2) Fill the graduate to the 50-ml mark with ethanol and mix thoroughly. Transfer 5 ml of the solution to an Erlenmeyer flask and put 5 ml of the ethanol into another. Treat both samples in exactly the same manner as described under CALIBRATION. Measure the optical density of the sample against the blank and determine the cyclopentadiene content from the calibration curve.

(3) Calculate the cyclopentadiene in the sample by the equation:

$$C = \frac{C' \times 10 \times 100}{W}$$

where

C′ is the weight in grams of cyclopentadiene in the test solution, as found from the curve.
W is the weight of the sample, in grams.
C is the weight per cent of cyclopentadiene in the sample under test.

DETERMINATION OF THE TOTAL CONJUGATED DIOLEFIN CONTENT OF ISOPRENE

This method is based on similar ones appearing in the *Laboratory Manual* of the Office of the Rubber Reserve and in the manual of ASTM D–2. It is primarily applicable to mixtures of C_5 hydrocarbons containing about 95 per cent of conjugated diolefins. Low results may be obtained if the *cis*-isomer of pentadiene-1,3 is present, since the reaction with this compound is not quantitative.

A weighed portion of the sample is put into the thoroughly chilled container shown in Figure 1, and the container attached to the apparatus. The container is heated and, by means of a stream of carbon dioxide, the gas is passed into the absorber and through a column of molten maleic anhydride. In this column, all of the conjugated diolefins will be absorbed in the reagent, while the other hydrocarbons and the carbon dioxide pass out at the top. The exit gas is then bubbled through a strong solution of caustic potash, where the carbon dioxide is absorbed. The unabsorbed gases are collected in a burette where the exact volume is noted.

Apparatus

(1) *Absorbing System*—The complete assembly, shown in Figure 1, consists of the following major items:

(a) Carbon dioxide cylinder, fitted with a two-stage reducing valve and attached to a rotameter.

(b) Sample bulb, constructed as shown in Figure 2. The stopcocks should be fashioned so that the carbon dioxide may either be passed through the sample or through the connecting line across the top. Provision must be made for closing any openings while this bulb is removed from the system. This is conveniently done by a third stopcock, as shown in Figure 2, and a pinch clamp on the line from the rotameter.

(c) The absorber (Figure 3) consists of three major pieces. The inner tube has hollow walls, so that the incoming gas may be directed to the bottom and then forced upward through the

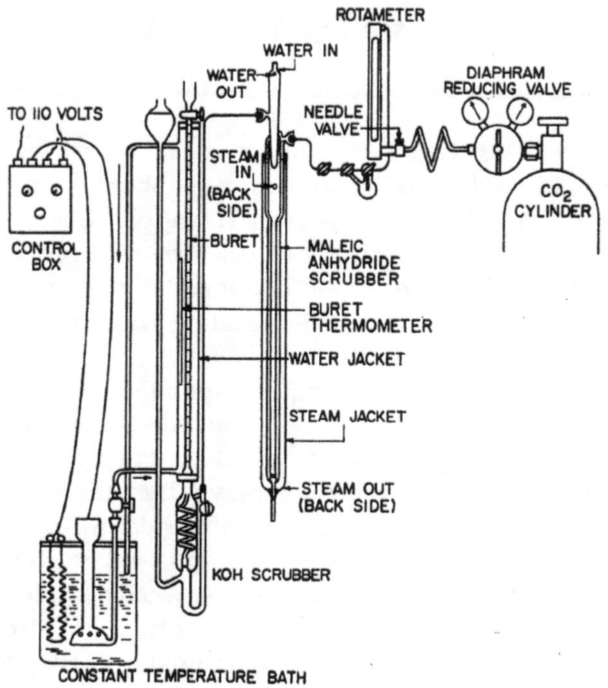

FIGURE 1. Apparatus for the Determination of Conjugated Diolefins in Isoprene

central portion which is filled with maleic anhydride. This tube is mounted in a jacket through which live steam can be circulated to keep the anhydride in molten condition. A small water-cooled condenser is mounted in the mouth of the tube to cool the gases before passage to the next unit.

(d) The scrubbing unit is filled with a strong solution of caustic potash. Attached to the top of the scrubber is a burette of 30 ml total capacity and graduated in 0.1-ml units. The burette is surrounded by a water jacket, and provision must be made for circulating water at a temperature of 150° ± 0.5°F through the jacket. A leveling bulb, also filled with caustic potash, is attached to the scrubber.

Satisfactory assemblies of the absorber (c) and the scrubber and burette (d) may be obtained from Podbielniak, Inc., 341 East Ohio Street, Chicago 11, Ill. The sample bulb is not a regular stock item, and must be made on special order. The constant-temperature bath may be prepared in any convenient manner.

Reagents

(1) *Maleic Anhydride,* minimum melting point, 52°C.

(2) *Di-*n*-Amylamine.*

(3) *Potassium Hydroxide Solution*—Dissolve 5 lb of the chemically pure reagent in 2.4 liters of distilled water. Store in a tightly stoppered bottle made of chemically resistant glass.

(4) *Stopcock Lubricant*—Mix 100 g of anhydrous glycerol, 29 g of starch, and 15 g of chemically pure *d*-mannitol. Heat slowly to boiling and cool, with frequent stirring.

Prepare the apparatus as follows:

(1) A cylinder of carbon dioxide will normally contain appreciable amounts of contaminants which will affect the results. To remove these impurities, quickly vent off some 50 to 75 per cent of the contents of a cylinder and use the remaining gas for this test.

(2) Assemble the rest of the apparatus as shown in Figure 1. During the preliminary work, do not include the sample bulb in the assembly and attach the discharge line from the rotameter to the stopcock near the absorber. Grease all of the joints lightly with the special stopcock lubricant.

(3) Weigh 65 g of maleic anhydride and 0.6 ml of di-*n*-amylamine into a flask. Attach a drying tube and heat until the anhydride has melted. Remove the condenser from the absorber and pour the liquid down the top. Replace the condenser and turn on the condensing water. Apply steam to the outer jacket.

Certain nondiolefinic hydrocarbons will polymerize in the presence of maleic acid, causing low results. To prevent this polymerization, the di-*n*-amylamine is added to the maleic anhydride. In addition, the mixture must be protected at all times from atmospheric moisture.

(4) Put about $\frac{1}{2}$ in. of mercury into the bottom of the scrubber under the burette. The mercury must be added through the side arm of the stopcock near the bottom. Fill the constant-

temperature bath with water, turn on the heat and controls, and heat the water to 150° ± 0.5°F. Turn on the circulator and circulate the water through the jacket surrounding the burette.

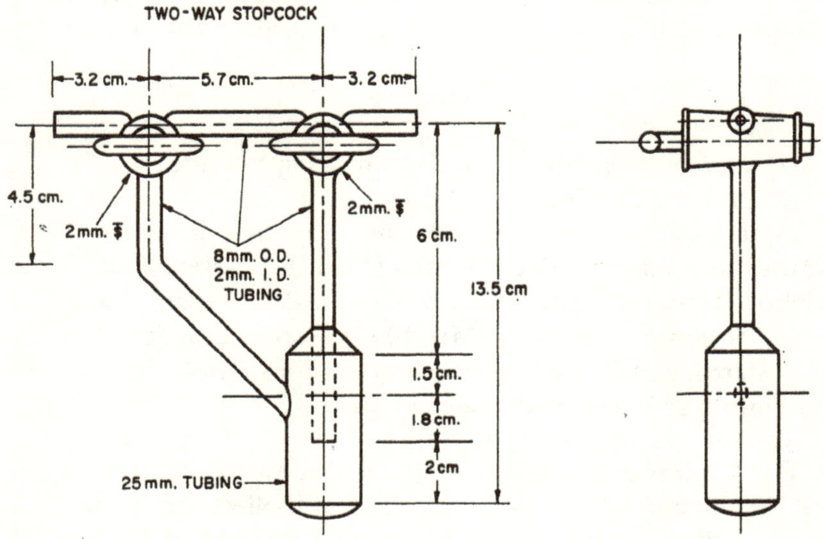

FIGURE 2. Sample Bulb for Vaporizing C_5 Hydrocarbons

(5) When the apparatus is completely assembled, fill the leveling bulb with the caustic potash solution, locating it near the bottom of the burette. Turn on the carbon dioxide and adjust the rate of flow to 30 ± 10 ml per minute, allowing the gas to leave the apparatus through the stopcock at the top of the burette. Flush the assembly for about 5 minutes to remove the air originally present and then fill the burette completely with the KOH solution. Allow the carbon dioxide to flow into the system at the prescribed rate for 30 minutes. At the end of 30 minutes, if the volume of gas remaining in the burette does not exceed 0.3 ml, the apparatus is ready for use.

Procedure

(1) Attach the sample bulb to the house vacuum line and evacuate it as completely as possible. Close all stopcocks, detach the bulb from the line, and weigh it to the nearest milligram.

(2) Attach the bulb to the bomb containing the sample by a short length of rubber tubing and immerse the bulb in a bath of liquid nitrogen. When thoroughly chilled, open the valve and allow 1 to 2 ml of the liquid sample to enter. Close the valve on the sample container and allow the tubing to drain for a few moments before closing the stopcock on the bulb. (This prevents removal of the grease by the sample.)

(3) Disconnect the sample bulb and immerse it in the nitrogen bath. When the sample is frozen, again evacuate the bulb to remove any air. Close all stopcocks, allow the bulb to come to room temperature, wipe the outside with a clean cloth, and weigh again. Record the weight of the sample.

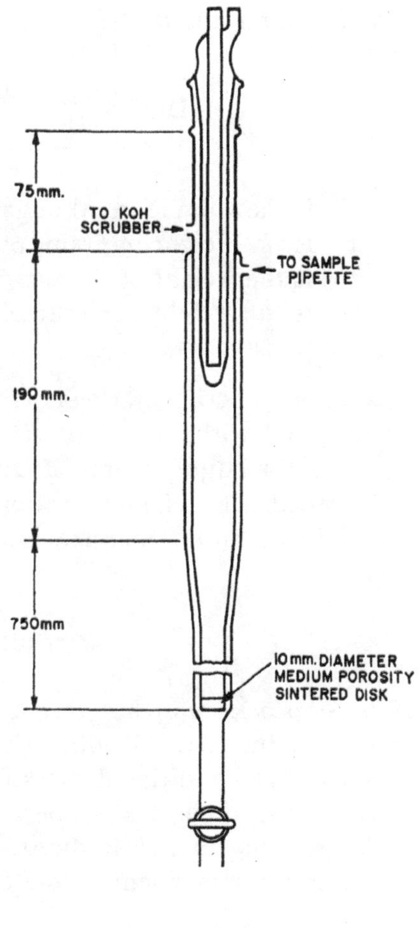

FIGURE 3. Maleic Anhydride Absorber

(4) Put the sample bulb into place in the apparatus and turn the stopcocks so that the carbon dioxide will pass across the top. Sweep out any air in the system, adjusting the rate of flow of the CO_2 to 30 ± 10 ml per minute. During this time, the apparatus must be ready for use under the prescribed conditions, i.e., steam must be flowing through the outer jacket of the absorber, condensing water through the condenser, and water at a temperature of 150°F circulating around the burette.

(5) When everything is ready, turn the stopcocks on the sample bulb so that carbon dioxide will pass through the sample. Allow the flow to continue until the sample is completely vaporized, warming the bulb if necessary. Continue until the volume of gas in the burette is constant over a 15 minute period. Note and record the volume of gas in the burette.

(6) Calculate the diolefin content of the sample by the equation:

$$D = 100 - \frac{V \times F \times 100}{W}$$

where

V is the volume in ml of gas remaining in the burette.
F is the weight in grams of 1 ml of this gas.
W is the weight of the sample in grams.
D is the weight per cent of diolefins in the sample under test.

Based on experimental work, the value for F may be taken as 0.00204 g/ml without serious error. For referee work, it is suggested that a large portion of the sample under test is treated as described. A weighed portion of the residue can then be treated in a similar manner, and the value for F determined exactly.

Precautions

The stopcocks must be greased and the apparatus be clean at the start of the test. All joints should be checked for leaks with soapsuds. As an additional check for leaks, after the air originally present in the apparatus has been removed, and with the carbon dioxide still flowing, close the stopcock near the KOH absorber and, after a few moments, close the one after the sample bulb.

No motion of the maleic anhydride should result in either case.

The maleic anhydride should be changed after four determinations, or after a lapse of 3 days, whichever occurs first. Water must be rigidly excluded. The caustic potash solution should be changed for every sample.

During the test, the temperature of the maleic anhydride scrubber must be kept as close to 212°F as possible. The temperature of the water in the jacket around the burette must be maintained at 150° ± 0.5°F.

For samples containing at least 95 per cent of diolefin, duplicate results by the same operator should not differ from one another by more than 0.2 per cent. For different operators and apparatus, the limit may be doubled.

The stopcock lubricant listed under REAGENTS is satisfactory for general use in this apparatus. There are certain substitutes available in the trade which may be used if desired. For the sample bulb, a suitable lubricant is *Cello-Grease,* available from the Fisher Scientific Co. under their catalog number 14–637. For the burette stopcock, a *Cello-Seal* is available from the same source under 14–636.

The procedure is primarily applicable to samples containing about 95 per cent diolefins. If the sample contains less than this amount, it is permissible to substitute a 110-ml burette for the 30-ml item listed under APPARATUS, and to examine a somewhat larger sample.

L.I.C. No. 200.52

COMPLETE ANALYSIS OF DISTILLED ISOBUTYLENE BY INFRARED SPECTROMETRY

This method was developed primarily to determine the butene-2 content of isobutylene feed stocks, but may be used for a complete analysis of samples of high purity. Although the method follows conventional infrared procedure, it does contain certain new features by which the precision and accuracy can be improved. By careful attention to details, each component can be determined if present in only a few hundreds of 1 per cent.

The method is intended for the examination of distilled isobutylene of at least 99 per cent purity. Prolylene, butadiene, and C_5 olefins will interfere, and should be either corrected for or removed.

The various C_4 components will show unique absorption peaks in the wavelength region of 5 to 15 μ. The exact locations of these peaks are determined by the examination of pure hydrocarbons, and the absorptivity of each determined. A standard sample is then prepared, containing the same impurities as the sample under test and in about the same amounts. This sample and the unknown are examined under the same testing conditions. The amount of each component in the unknown can be determined by comparing the behavior of the known and the unknown samples under identical conditions.

The observed absorption of all compounds will vary with the pressure inside the cell of the instrument. To minimize possible errors from this source, all examinations are made at three different pressures, the absorptivity at each calculated to that at 100 mm, and the averages used in the final calculations.

To detect the presence of interfering compounds, the sample under test and the standard sample should be scanned over the critical area, and the two patterns compared. Any marked difference in the general patterns will indicate the presence of some interference.

Apparatus

(1) *Spectrometer*—The Beckman model IR–2 was used in the development of this method. Any other type may be used, providing that it is equipped with a gas cell, at least 37 cm in length, gas-handling equipment, a vacuum pump, and a constant-temperature circulating bath.

The instrument should be located in a room where the temperature does not vary by more than 4 degrees. Greater variations in temperature will cause shifts in the wavelength of the instrument and seriously affect the results.

Reagents

(1) *Pure Hydrocarbons*—A number of pure compounds for calibration purposes are available from the Phillips Petroleum Company, Bartlesville, Okla. The complete set for this method should include butene-1, butene-2, (*cis* and *trans*), isobutylene, butane (*iso*- and *normal*), butadiene-1,3 and any C_3 or C_5 hydrocarbon which might be present.

(2) *Standard Sample* containing the same components in essentially the same amounts as the sample under test.

Sampling

Samples should be collected and handled in standard containers. Whenever it is to be shipped outside the refinery, the container must meet all the requirements of the Interstate Commerce Commission. As a safety measure, the amount of sample should be kept at a minimum, and under no circumstances should the volume of the liquid sample fill the container more than two thirds full.

The portion of the sample which is to be examined must not contain free water. If necessary, a drying tower filled with caustic pellets or Ascarite should be located between the sample container and the instrument.

Calibration

(1) Assemble the apparatus and prepare for testing in accordance with the manufacturer's instructions.

(2) Evacuate the cell completely and introduce a portion of the pure butene-1 until the pressure inside the cell is about 500 mm. Scan in the range of 5 to 6 μ, and find the exact location of the peak height at or near 5.41, and the best slit width for this compound.

Butene-1 absorbs sharply in this range and the exact wavelength setting is critical. When the location is found, it is suggested that some type of permanent mark is made on the instrument, so that the dial can always be returned to that same setting. When developing the method, a slit width of 0.20 mm was found satisfactory.

(3) When these conditions have been established, evacuate the cell again and adjust the amplification control until the absorbancy (optical density) of the empty cell at observed conditions of wavelength and slit width is zero. From here on, do not change the amplification.

(4) Again introduce the butene-1 until the pressure is 500 mm. Note and record the exact pressure in the cell and the absorbancy of the gas under the optimum conditions. Repeat these observations at pressures of 450 and at 400 mm, recording in each case the exact pressure and the absorbancy.

(5) Empty the cell completely and introduce another of the standards until the pressure is 600 mm. Without changing the amplification, adjust the other controls until the optimum conditions as to peak height and slit width are found. Note the absorbancy and the slit width under these conditions. Without changing the slit width, repeat these observations at pressures of 550 and 500 mm, and finally on the empty cell.

While the exact conditions will vary with the instrument, the data listed in Table 1 were noted in the development of this method.

As previously stated, butene-1 absorbs more strongly than any of the others. Therefore, as the others are examined, it is necessary to increase the pressure (observations are made at 600, 550, and 500 mm, rather than 500, 450, and 400 mm) and the slit width. The absorbancy of the empty cell will increase as the slit is widened, so that this observation must be repeated for each compound. However, the amplification control must not be changed after the original setting for butene-1 was made.

TABLE 1

Compound	Wavelength	Slit Width
Isobutane	8.45	0.50
trans-Butene-2	10.35	0.72
n-Butane	13.33	1.40
cis-Butene-2	14.77	3.00

(6) Repeat this procedure on each of the remaining pure compounds, finding first the optimum conditions as to slit width and wavelength, and then noting the absorbancy at pressures of 600, 550, 500 mm, and finally on the empty cell.

Procedure

(1) Evacuate the cell completely, set the controls for testing butene-1 (i.e., wavelength at the critical setting near 5.41, and slit width at or near 0.20) and adjust the amplification control until the absorbancy of the empty cell is zero. From here on, do not change the amplification control.

(2) Introduce the standard sample until the pressure is 500 mm, and note the absorbancy. Repeat the reading at pressures of 450 and 400 mm.

(3) Set the dials for examination of isobutane (i.e., 8.45 μ, width 0.50 mm), introduce the standard sample until the pressure is 600 mm, and note the absorbancy. Repeat at pressures of 550 and 500 mm, and on the empty cell, recording in each case the exact pressure and the absorbancy. Repeat the entire procedure under proper conditions for each of the other components of the standard sample.

(4) Empty the cell completely and repeat the entire procedure with the unknown sample, making the observations at the lower pressures (500, 450 and 400 mm) when testing for butene-1 and at the higher pressures for each of the others. Include the ab-

sorbancy of the empty cell at each setting of wavelength and slit width.

Calculations

(1) Isobutylene does not follow the gas laws perfectly, so that corrections for this deviation must be applied. The corrected values for various observed pressures are shown in Table 2. Using this table, determine the corrected pressure for *all* recorded values for observed pressure.

(2) Beginning with the calibration compounds, determine the absorptivity of each by the equation:

$$a = \frac{(A_S - A_O) \times 100}{Pc}$$

where

A_S is the observed absorbancy (optical density) of the calibration compound at its key setting.

A_O is the absorbancy of the empty cell at the same setting.

Pc is the corrected pressure.

a is the absorptivity of the compound under test at that particular setting.

Repeat this calculation for each compound, calculating the absorptivity at the key setting only. (Note that the value for A_O is different for each compound.) Record the absorptivity as a_3 for each compound.

(3) Using the same equation, calculate the absorptivity of the sample under test at each key wavelength setting. Record the absorptivity at each setting as a_2. Repeat the calculation for the reference sample at each key setting, and record these values as a_1.

(4) Calculate the amount of each component in the unknown by the equation:

$$C = \frac{(a_1 - a_2)}{a_3}$$

where

the "a" values are as previously described.

C is the correction which is to be added to the amount of each component in the reference sample to obtain the corresponding amount in the sample under test. (Sometimes the C value will be a minus quantity; in this case, it is to be subtracted from the corresponding amount in the reference sample.)

(5) Add together the amounts of the individual contaminants, subtract the sum from 100 per cent, and consider the difference as the content of isobutylene in the sample.

TABLE 2

Observed Pressure, millimeters	Corrected Pressure, millimeters	Observed Pressure, millimeters	Corrected Pressure, millimeters
10	10.0	310	313.8
20	20.0	320	324.0
30	30.1	330	334.3
40	40.1	340	344.5
50	50.1	350	354.8
60	60.2	360	365.1
70	70.2	370	375.4
80	80.3	380	385.7
90	90.3	390	396.0
100	100.4	400	406.3
110	110.5	410	416.6
120	120.6	420	427.0
130	130.6	430	437.3
140	140.7	440	447.6
150	150.9	450	458.0
160	161.0	460	468.4
170	171.1	470	478.7
180	181.3	480	489.1
190	191.4	490	499.5
200	201.6	500	509.9
210	211.8	510	520.2
220	221.9	520	530.6
230	232.1	530	541.1
240	242.3	540	551.5
250	252.5	550	562.0
260	262.7	560	572.4
270	272.9	570	582.8
280	283.1	580	593.2
290	293.3	590	602.7
300	303.6	600	614.2

Typical Calculation

Suppose that in the calibration examination of *cis*-butene-2 the following results were obtained:

Slit Width	14.77 μ
Cell Thickness	3.00 mm
Absorbancy of empty cell	0.15 units
Absorptivity of the compound (Average a_3 value)	0.06

Let us assume that under the same conditions the results listed in Table 3 were obtained:

TABLE 3

	Reference Sample				Unknown Sample		
Observed Pressure	Corrected Pressure	Observed A	Net A	a Value	Observed A	Net A	a Value
600.0	614.2	0.203	0.053	0.0086	0.193	0.043	0.0070
500.0	509.9	0.194	0.044	0.0086	0.185	0.035	0.0068
400.0	406.3	0.185	0.035	0.0086	0.178	0.028	0.0068

First, change the observed pressure, as shown in the first column of Table 3, to the corrected pressure, as shown in the second. From then on:

For the reference sample, subtract the correction for the cell absorbancy from the observed absorbancy, and enter the difference as the net absorbancy at that pressure.

$$0.203 - 0.150 = 0.053$$
$$0.194 - 0.150 = 0.044$$
$$0.185 - 0.150 = 0.035$$

Multiply the net absorbancy by 100, and divide by the corrected pressure, in millimeters. Record as a_2.

$$5.3/614.2 = 0.0086$$
$$4.4/509.9 = 0.0086$$
$$3.5/406.3 = 0.0086$$

Obtain an average value for a_2.

0.0086

Repeat the entire procedure on the *unknown* and obtain an average "a" value (a_1).

0.0068

Subtract a_2 from a_1, and note whether this yields a plus or minus quantity.

$$\begin{array}{r} 0.0068 \\ -0.0086 \\ \hline -0.0018 \end{array}$$

Divide the difference by the "a" value for the pure compound being determined at this setting (a_3).

$$-0.0018/0.060 = -0.03\%$$

Consider the difference as the correction to be added to the amount of that particular component in the reference sample to obtain the corresponding amount in the unknown.

Reference Sample contains	0.07%
Correction	−0.03
	0.04%

The sample under test would, therefore, contain 0.04 per cent of *cis*-butene-2. Repeat for each of the other components. The final analysis would be similar to the data in Table 4.

TABLE 4

Compound	Reference Sample Contains	Correction	Unknown Sample Contains
	%		%
cis-Butene-2	0.07	−0.03	0.04
trans-Butene-2	0.07	−0.07	0.00
normal Butane	0.00	0.00	0.00
Isobutane	0.10	+0.08	0.18
Butene-1	0.37	−0.04	0.33
Isobutylene	99.39		99.45

Normally, butadiene is not present in samples of isobutylene. However, if present it will interfere sharply, since it absorbs strongly between 5.4 and 5.5. If the presence of butadiene is suspected, particular attention should be paid to the absorption pattern in this range.

At the discretion of the operator, the directions outlined under PROCEDURE may be modified as follows: Zero the instrument as described in Paragraph 1. Set the controls to examine *cis*-butene-2 (i.e., 14.77 and 3.00), fill the cell with the standard sample to a pressure of 600 mm, and examine for all compounds *except* butene-1, rotating the dial for each of the others. After finishing with isobutane, reduce the pressure to 550 mm, and gradually return to the other end of the dial. Reduce the pressure to 500 mm, and work back again, this time including butene-1. Without changing the critical setting at 5.41, reduce the pressure to 450 and then to 400 mm. Empty the cell, replace with the unknown and, still not changing the settings, obtain readings for butene-1 at the three pressures. When the last has been obtained, finish as usual. If done in this manner, the wavelength dial need be set only once for ALL measurements at the critical setting near 5.41.

Any type of infrared spectrometer may be used for carrying out this method. Some models have a series of terret stop pins, so that the wavelength dial may be returned automatically to any given setting. Obviously, with such instruments it will not be necessary to make any marks on the dial.

L.I.C. No. 220.10

SOLUBILITY OF PENTACHLOROPHENOL IN PETROLEUM OILS

Certain chemical and wood-preserving companies require an estimation of the solubility of pentachlorophenol in petroleum oils to be used for specific purposes. While there is no generally accepted method for this determination, it is believed that the method outlined here is the most widely used. This method was developed by the Chapman Chemical Co.

Reagents

Pentachlorophenol—This should be the commercial grade meeting Federal specification TT-W-570. A satisfactory product may be purchased from the Chapman Chemical Co., Memphis, Tenn., or the Wood Treating Chemicals Co., St. Louis, Mo.

Procedure

(1) A 45-g sample and 5 g of pentachlorophenol are mixed and stirred to complete solution. If necessary, the mixture may be heated, but in no case should the temperature be above 180°F, or high enough to cause appreciable "smoking."

Usually there will be a small but visible amount of undissolved matter, which consists of impurities in the reagent. This small residue may be ignored.

(2) After mixing, the blend is stored undisturbed for 16 hours at 75° ± 3°F. After storage, the blend is examined for separation of crystals of pentachlorophenol. If there is any such separation, this is an indication that the solubility of pentachlorophenol has been exceeded.

On dark, heavy oils the blend may be examined by transmitted light, or decanted through a coarse filter paper. Any solidification at room temperature indicates heavy crystal formation.

The undissolved impurities will be in the form of a flocculent precipitate, which will wash freely from the sides of the container. Such a precipitate may be ignored; only the crystalline formation of the solute is of importance. The pentachlorophenol will be in the shape of needlelike crystals, easily distinguished from the flocculent precipitate.

(3) If there is no visible evidence of solid crystals, the solvency of the oil under test is reported as *above 10 per cent.* If there is any visible separation of the crystals, the solvency is reported as *below 10 per cent.*

L.I.C. No. 230.01A

DETERMINATION OF LOWER KETONES

This method was developed by the Chemical Products Laboratory from a similar test appearing in the literature as a means of accurately measuring the purity of methyl ethyl ketone. The procedure is based on the fact that compounds having a carbonyl group will react with hydroxylamine hydrochloride as follows:

$$\begin{matrix} R_1 \\ \quad\diagdown \\ \quad\quad CO \\ \quad\diagup \\ R_2 \end{matrix} + \begin{matrix} H \\ \quad\diagdown \\ \quad\quad NOH{\cdot}HCl \\ \quad\diagup \\ H \end{matrix} = \begin{matrix} R_1 \\ \quad\diagdown \\ \quad\quad C{=}NOH \\ \quad\diagup \\ R_2 \end{matrix} + HCl + H_2O$$

Theoretically, one mol of acid will be formed for every mol of ketone originally present. However, the reaction does not quite go to completion, and a small correction factor must be applied, as is outlined under CALCULATIONS. Formerly the amount of liberated acid was measured by volumetric titration, using methyl orange xylene cyanole as the indicator. The end point in this titration is quite difficult to see, and even in the hands of a skilled operator, the error may be in the range of 0.25 to 0.70 per cent. In this version, the estimation of the liberated acid is made by the use of a *p*H meter, so that the range of error is only 0.03 to 0.10 per cent.

This method is applicable to samples containing methyl ethyl ketone or acetone. It may also be used for the determination of other ketones, providing that the proper correction factors, as outlined under CALCULATIONS, are determined experimentally. The method may not be used in the presence of aldehydes, or on samples containing a mixture of more than one ketone.

Any free acid definitely interferes. However, the amount of this free acidity can be determined on a separate sample and a correction made.

Apparatus

(1) p*H Meter*—The recommended type of instrument is the Cambridge electron-ray *p*H machine, with a sensitivity of 0.005

*p*H. Any other type of instrument may be used, if the readings are accurate to at least 0.01 *p*H.

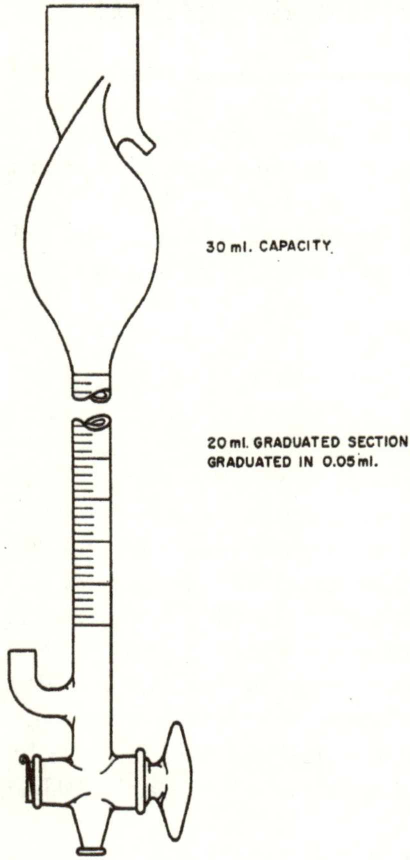

FIGURE 1. Semiautomatic Burette

(2) *Burette*—For the most precise work, the burette should be graduated in increments of 0.05 ml. The recommended burette is of the semiautomatic delivery type, with a bulb near the top of exactly 30 ml, and a stem of 20 ml capacity, graduated in 0.05 ml. The burette should be calibrated by weighing increments of freshly distilled water. This calibration must be carefully done, since an error of 0.01 ml in the titration appears, as an error of 0.02 per cent in the final answer.

(3) *Weighing bottles*—Since ketones, in general, are rather volatile, extreme care must be taken in weighing the sample for analysis. The usual weighing bottle, equipped with a glass stopper and a ground-glass joint, is satisfactory, if the bottle is

placed near the surface of the liquid in the titration flask before opening. A thin-walled glass ampoule, which is also broken while under the surface of the titration liquid, is also acceptable.

Reagents

(1) *Sodium Hydroxide, 03 N*—This solution must be as free as possible from carbonates. In 450 ml of recently boiled and cooled distilled water 454 g of reagent-grade caustic soda is dissolved and stored in a bottle provided with a soda-lime guard tube. After standing for a few days, 15.6 ml of the clear supernatant liquid is diluted with freshly boiled and cooled distilled water to a total of 1000 ml and stored in a bottle which is also protected by a soda-lime guard tube. The final solution must be accurately standardized, since an error of 0.0001 in the normality will result in a difference of 0.02 per cent in the final answer. To standardize the solution, Bureau of Standards grade of benzoic acid or potassium acid phthalate may be used. If benzoic acid is used, 1.6 g of the acid is weighed into a weighing bottle, and the bottle placed into a 250-ml Erlenmeyer flask containing 25 ml of neutral alcohol. After the acid has dissolved, 75 ml of water is added, the liquids heated to boiling, and titrated with the caustic, using phenolphthalein as the indicator. If acid potassium phthalate is used, the reagent is dried to constant weight at 105°C and 2.8 g is dissolved in 100 ml of distilled water, heated to boiling, and titrated hot. Whichever reagent is used, at least three determinations should be made, or standardization continued until the results agree to within ±0.0002 N.

(2) *Indicator*—The indicator is methyl orange xylene cyanole, which may be obtained from the Eastman Kodak Company, Rochester, N. Y., in the form of a solution having a *p*H range of 3.2 to 4.2.

(3) *Hydroxylamine Hydrochloride*—Since the reagent as purchased is seldom 100 per cent pure, and since some of the impurity is hydrochloric acid, a special solution must be prepared for this test. It is imperative that the directions given are followed exactly.

25 g of chemically pure hydroxylamine hydrochloride is dissolved in distilled water to make 250 ml, and 12 drops of methyl orange xylene cyanole indicator added. For best results, this

solution should be prepared fresh each day. From this solution, 25 ml is pipetted into a 500-ml Erlenmeyer flask, and 190 ml of distilled water added. The diluted solution is then titrated with the 0.3 N NaOH solution until the *p*H is between 3.85 and 3.95, as measured by the *p*H machine. The exact volume of NaOH is noted and recorded. Just before weighing the sample of the ketone to be analyzed, exactly nine times the amount of NaOH solution used to titrate the aliquot portion of the hydroxylamine solution is added to the remainder of the solution in the volumetric flask and mixed thoroughly.

Procedure—Methyl Ethyl Ketone

(1) The size of the sample to be weighed should be such that the final sample will contain 1 ±0.05 g. For methyl ethyl ketone of over 95 per cent purity, 1 g will be sufficient; if the purity is below 95 per cent, a larger amount must be taken. The sample is weighed to 0.1 mg in one of the containers described under APPARATUS.

(2) A blank is prepared by pipetting 25 ml of the neutralized hydroxylamine solution as under REAGENTS into a 500-ml Erlenmeyer flask containing 190 ml of distilled water. This solution should have a steel-gray color; if it is not so colored, the solution has not been properly neutralized. The *p*H of this steel-gray colored solution is measured and recorded. The observed *p*H *must* lie between 3.85 and 3.95, preferably above 3.90.

(3) Into a second 500-ml Erlenmeyer flask, are placed 150 ml * of distilled water, 25 ml of the neutralized hydroxylamine solution, and the weighed sample. After thorough mixing, the sample is titrated with the 0.3 N NaOH solution until the color matches the steel-gray color of the blank. The *p*H of the sample is then noted and recorded. The volume of the caustic solution required to reach this color is also recorded.

Calculations

(1) It will seldom be found that the *p*H of the solution at the

* Only 150 ml of water is placed in the Erlenmeyer flask, instead of the 190 added for the blank. Since about 40 ml of caustic solution will be required for the titration, when the end point is reached, the volume will then be very nearly the same as that of the blank.

end of the titration is exactly the same as that of the blank. The error may be calculated by the following formula:

$$\frac{0.7140 \times (pH \text{ of Blank} - pH \text{ of Sample})}{\text{Normality of NaOH}} = \text{milliliters of error}$$

If the *p*H of the sample is lower than that of the blank, the sample was undertitrated, and the calculated error must be added to the observed volume of caustic. If the *p*H of the sample is higher than that of the blank, the sample was overtitrated, and the calculated error must be subtracted from the observed volume.

(2) The percentage of methyl ethyl ketone may then be calculated as follows:

$$\% \text{ MEK} = \frac{(\text{ml NaOH} \pm \text{ml of error}) \times \text{N of NaOH} \times 7.210}{\text{Wt. of Sample} \times 0.9800}$$

(3) All calculations should be carried out to the fourth significant figure. Generally, slide-rule calculations are not sufficiently accurate. The figure 0.9800 is used to compensate for the incomplete oximation of the ketone.

Procedure—Acetone

(1) The apparatus, reagents, and procedure for acetone are all the same as for methyl ethyl ketone, with the single exception that the size of sample taken for analysis is 0.8 g, instead of 1.0 g.

(2) The method of calculation for acetone is the same as for methyl ethyl ketone, but the factors are slightly different, as follows:

$$\frac{0.7851 \times (pH \text{ of Blank} - pH \text{ of Sample})}{\text{Normality of NaOH}} = \text{milliliters of error}$$

$$\% \text{ Acetone} = \frac{(\text{ml of NaOH} \pm \text{ml of error}) \times \text{N of NaOH} \times 5.808}{\text{Wt. of Sample} \times 0.9855}$$

Cautions

(1) Other ketones and aldehydes will react in a similar fashion with hydroxylamine hydrochloride. This method, therefore, cannot be used on samples containing more than one ketone or/and aldehyde.

(2) Alcohol, water, hydrocarbons, and ethers if present in concentrations under 10 per cent will not affect the accuracy seri-

ously. Any acid present in the sample will cause an error in the titration. However, this may be corrected by determining the acid on a separate sample. This should be done by using the *p*H meter.

(3) The analyses should be carried out in a room whose temperature is close to 25°C. The directions given for preparation of reagents, *p*H of solutions, etc., must all be followed closely, since the factors used to compensate for incomplete oximation (0.9800 for methyl ethyl ketone and 0.9855 for acetone) are valid only for the conditions prescribed.

DETERMINATION OF THE ALCOHOL CONTENT BY ACETYLATION

When an alcohol reacts with acetic anhydride on the presence of pyridine the titratable acidity of the reaction mixture is reduced as follows:

$$(CH_3CO)_2O + H_2O \longrightarrow 2\,CH_3COOH$$
$$(CH_3CO)_2O + ROH \longrightarrow CH_3COOR + CH_3COOH$$

The function of the pyridine is to act as a diluent and to combine with the free acetic acid liberated in the reaction, thus aiding the completion of the titration. The pyridine acetate so formed is completely titratable as the free acetic acid equivalent.

This method has been adapted from a smilar one appearing in the literature. It is applicable to specification-grade alcohols of low molecular weight. For the analysis of mixtures of different alcohols, or for dilute solutions, refer to section 240.10A.

Olefins, esters, and aromatic hydrocarbons do not interfere, nor do aldehydes unless present in concentrations greater than 10 per cent. Mercaptans, amines, and some ethers interfere. Since any free acidity or alkalinity will introduce an error, such acidity or alkalinity must be determined on a separate portion and the proper correction applied to the final result.

Apparatus

(1) *Weighing Bottle*—Should be glass-stoppered, measuring about 15 × 50 mm.

(2) *Burette* of automatic type, with a bulb at the top having a capacity of 60 ml. When not in use, precautions should be taken to keep the stopcock from sticking.

(3) *Reflux Assembly* consisting of a 20-in. Allihn condenser and a 250-ml Erlenmeyer flask, fitted together with a ground-glass joint. This assembly is available at the Corning Glass Works under the condenser as item 2440 and the flask as item 5000.

Reagents

(1) *Pyridine*—Store the reagent-grade liquid in an amber bottle over caustic soda and filter through dry paper just before use.

(2) *Acetylating Reagent*—Filter 880 ml of the pyridine, dried over caustic soda, into an amber-colored bottle and add 120 ml of chemically pure acetic anhydride. Store the bottle in a dark place when not in use.

(3) *Standard Sodium Hydroxide Solutions*—Two solutions will be needed, one 0.8 N and the other 0.01 N. It is important that both solutions are free of carbonates. To prepare the approximately normal solution, dissolve 1 lb of chemically pure sodium hydroxide in 1 pt of distilled water and allow the resulting solution to stand for several days in a tightly stoppered container. At the end of at least 1 week, decant or pipette 42 ml of the clear supernatant liquid into a 1-liter volumetric flask and fill it to the mark with recently boiled and cooled distilled water. Standardize against either potassium acid phthalate or benzoic acid and store the final solution, protecting it from carbon dioxide by means of a soda-lime guard tube. To prepare the 0.01 N solution, measure an appropriate amount of the stronger solution into a 1-liter volumetric flask and fill to the mark with recently boiled and cooled distilled water. Protect this solution also by a soda-lime guard tube.

To standardize the stronger solution, weigh accurately about 6 g of potassium acid phthalate (or 4 g of benzoic acid) into an Erlenmeyer flask. Add some 200 ml of carbonate-free water when the phthalate is used, or 50 ml of neutralized alcohol followed by 150 ml of water when benzoic acid is used. In either case add a few drop of phenolphthalein indicator, and titrate to a sharp pink end point, completing the titration in a boiling solution. At least three such determinations should be made. Calculate the normality by either of the equations:

$$\text{Normality} = \frac{\text{Weight of phthalate}}{0.20422 \times \text{milliliters of solution}}$$

or

$$\frac{\text{Weight of benzoic acid}}{0.12205 \times \text{milliliters of solution}}$$

(4) *Hydrochloric Acid, 0.1 N*—Mix 9 ml of the concentrated acid with sufficient distilled water to make 1 liter. Standardize against the standard caustic solution.

(5) *Indicator*—Dissolve 0.1 g of phenolphthalein in 100 ml of neutralized 91 per cent isopropyl alcohol.

Procedure

(1) Weigh about 1 g of the sample into each of two tared weighing bottles and determine the weight of each to the nearest 0.1 mg. To avoid evaporation loss, keep the liquid away from the ground-glass joint and close the bottle as quickly as possible.

This procedure is satisfactory for ethyl and propyl alcohols. When testing butyl or higher-molecular-weight alcohols, increase the weight of the sample proportionately. When testing methyl alcohol, weigh the sample in a sealed glass ampoule.

(2) Measure 25 ml of the acetylating reagent into each of three Erlenmeyer flasks. Reserve one of these flasks as a blank and add 2 drops of water. Put a weighed sample into each of the other two and connect all three to reflux condensers. Boil the contents gently for 2 hours.

(3) During this time, weigh about 50 g of the sample into a clean flask, heat to boiling, add a few drops of indicator, and determine whether or not the sample is acid or alkaline. In either case, titrate with standard alkali or acid solution, as applicable, and record the necessary volume of the standard solution.

(4) At the end of the refluxing period cool the flasks to room temperature and rinse each condenser with about 50 ml of distilled water, collecting the rinsing in the proper flask. Titrate the contents of each flask to a sharp pink end point with the standard sodium hydroxide solution, recording the necessary volume in each case.

Calculations

(1) Determine the necessary correction for any acidity or alkalinity in the original sample by the equation:

$$C = \frac{V \times N \times 0.1}{S}$$

where

- V is the volume in milliliters of either standard acid or alkali needed to correct for any original acidity or alkalinity.
- N is the normality of this solution.
- S is the weight of this sample in grams.
- C is the correction to be added or subtracted. The value for C is obtained in centiequivalents, and should be subtracted when the original sample is alkaline and added when acid.

(2) Determine the alcohol content of the sample by the equation:

$$A = \frac{(B - T) \times N \times 0.1}{W} \pm C$$

where

- B is the volume in milliliters of the caustic solution required for the blank.
- T is the corresponding volume for the sample.
- W is the weight of this sample in grams.
- A is the alcohol content of the sample under test, expressed as centiequivalents per gram of sample. To determine the weight percentage of alcohol, multiply the value for A by the molecular weight of the alcohol under test. The molecular weights of the common alcohols are as follows:

Methyl	32.03	Propyl	60.06
Ethyl	46.05	Butyl	74.08
Amyl	88.09		

WATER DILUTION OF ALCOHOL

Two methods are outlined in this section. The first is used primarily for detecting the presence of water-insoluble impurities in a water-soluble solvent, while the second determines the solubility of a given product in varying amounts of water. Since the apparatus and procedure are essentially the same, these two methods are discussed together.

Apparatus

(1) *Cylinder*—A clear transparent glass cylinder, preferably glass stoppered, should be used. The exact size is immaterial, but the volume should be at least 100 ml and the height of the graduated portion, at least 7 in.

Procedure—Detection of Impurities

(1) The sample is brought to 20°C, and a suitable amount poured into the cylinder. The proper amount of distilled water, also at 20°C, added, the cylinder stoppered, and shaken vigorously.

The quantities of water and sample taken should be in the ratio dictated by the specifications and sufficient to fill the cylinder about seven eights full.

(2) The cylinder is allowed to stand undisturbed for a few moments, and then carefully examined for haze or turbidity. It cannot be too strongly emphasized that this examination must be carefully and conscientiously done. The liquid in the cylinder should be inspected in two planes: by looking down through the full depth of the liquid, and by looking through the sides of the cylinder. The examination should be made in soft, diffused light, and the cylinder held against a background which will readily permit the detection of any change in color.

(3) If there is no indication whatever of turbidity, cloudiness, or bluish haze, the sample should be reported as: *Passes.* If there

is any evidence of haze, the sample should be reported as: *Does not pass.*

Procedure—Miscibility

(1) Place 20 ml of the sample under test in the cylinder, and adjust the temperature to 20°C.

(2) Add 80 ml of distilled water in small amounts. The exact increments are shown in Table 1.

TABLE 1

First 5 ml	1 ml
5 to 20 ml	5 ml
20 to 80 ml	10 ml

(3) After each addition, the cylinder is stoppered, shaken, and after settling, examined for haze or cloudiness in the manner described previously. If there is no indication at any time of haze, the sample should be reported as: *Miscible in all proportions.* If there is at any time a sign of haze, the sample should be reported as: *Cloudy at .... per cent dilution.*

DETERMINATION OF ALCOHOL CONTENT BY SPECIFIC GRAVITY

A convenient means of measuring the purity of specification-grade water-soluble alcohols consists of determining the specific gravity and obtaining the purity by comparison with a set of tables. It is obviously important that the determination of the specific gravity is accurate. It is recommended that all determinations of specific gravity for this purpose should be made by the use of a Westphal balance. Experience has shown that when the gravity is determined by a hydrometer, the final alcohol content may be in error by as much as 0.1 volume per cent.

This method is suitable for the examination of plant alcohols (methyl, ethyl, and isopropyl) where the sample consists essentially of a mixture of water and one of the alcohols. It cannot be used on samples which contain more than minute amounts of other organic liquids or dissolved gases which cannot be removed by distillation. It may be used on samples containing suspended or dissolved heavy liquids or solids, providing that the sample is distilled and the alcohol content of the distillate calculated to that of the original sample.

Specific gravity is the relationship between the weight of a given volume of the sample under test at a given temperature and the weight of the same volume of distilled water at a given temperature.

The role of temperature in this relationship is important, and should be clearly understood. The specifications for an alcohol will usually list two temperatures such as 15°/15°C, or 15°/4°C. In each case, the first figure is the temperature to which the sample is to be adjusted before making the test, while the second is the temperature to which the water is to be adjusted. Thus, a specification of 15°/15°C means that the weight of a given volume of alcohol at 15°C is to be compared with the weight of the same volume of water, also at 15°C.

Apparatus

Westphal Balance available from any chemical supply house. Several different models are on the market, any one of which may be used. The balance consists of the usual beam, which is notched at regular intervals along one arm, a plummet, and a set of riders. The plummet will displace a certain amount of water which, in turn, is equivalent to the weight of the largest rider. The smaller riders are 0.1, 0.01, and 0.001 as heavy as the largest. By suspending the plummet in the liquid at the proper temperature and bringing the balance into equilibrium, the specific gravity may be read directly from the location of the riders.

FIGURE 1. Chain-O-Matic Balance

If the room temperature differs markedly from the test requirement, some provision must be made for controlling the temperature of the sample. This may be done in any convenient manner. The simplest arrangement consists of partly immersing the cylinder containing the sample in a water bath of carefully controlled temperature. A more elaborate assembly is shown in Figure 1, and consists essentially of a double-walled cylinder. Provision is made for circulating water at a carefully controlled temperature between the walls of the cylinder, and a stopcock at the bottom permits the removal of the sample after the test is finished.

Procedure

(1) Consult the specifications for the sample under test and determine the required ratio between test and calibration temperatures (15°/4°, 20°/20°, etc.).

When testing ethyl or isopropyl alcohol, the specifications normally read 20°/20°C. This means that both the sample and the water must be at 20°C when the measurements are made. However, the ratio for methyl alcohol may read 15°/4°C. This means that the balance must be brought into equilibrium with the water at 4°C, and the sample measurement made at 15°C.

(2) Set up the balance in a level place which is free of drafts and sudden changes in temperature. Attach the plummet to the arm and if necessary bring the balance into equilibrium by turning the setscrews at the bottom.

(3) Fill the sample container with freshly boiled and cooled distilled water, previously brought to the required temperature. Immerse the plummet in the water and put the container in place. Suspend the largest rider from the hook at the end of the arm. At this point, the balance should be in equilibrium. If it is not, turn the counterweight on the opposite end of the arm until the balance is in equilibrium.

There will be a small twist of wire immediately above the loop in the top of the plummet. The liquid level should be such that this twist will be completely immersed in the water, to avoid errors due to surface tension.

(4) Remove the sample container, discard the water, and rinse the container several times with the sample. Fill the container with the sample, put it under the arm, immerse the plummet, and put the largest rider on the hook at the end of the balance

arm. If the plummet sinks, move the rider and put it in the notch marked 0.9. If the plummet sinks again, put the rider on the notch marked 0.8, or farther back as necessary until the plummet rises. Then put the next-to-largest rider on the arm in the notch marked 0.9, and repeat. Continue this process, using the progressively smaller riders, until the balance is in equilibrium. Read the specific gravity from the location of the riders.

Example: If the largest rider is on the notch marked 0.7, the next-to-largest on 0.5, and the next on 0.3, the specific gravity is 0.753.

At all times the temperature of the sample must be within 0.1 °C of that required by the specifications.

(5) If necessary, determine the alcohol content of the sample from the attached tables and charts.

The directions outlined here presume the use of a balance which is equipped with riders of varying size. Some laboratories prefer the use of the Chain-O-Matic balance, as shown in Figure 1. This is permissible, but then the procedure must be modified accordingly. When using this type of balance, the first digit to the right of the decimal is obtained by the use of the rider, and the others from the vernier attached to the chain.

TABLE 1. Specific Gravity of Mixtures of Ethyl Alcohol and Water by Weight *

Per Cent Alcohol by Weight	15°C	20°C	25°C	Per Cent Alcohol by Weight	15°C	20°C	25°C
0	0.99913	0.99824	0.99708	51	0.91566	0.91164	0.90758
1	0.99725	0.99636	0.99521	52	0.91344	0.90940	0.90533
2	0.99543	0.99453	0.99338	53	0.91120	0.90715	0.90307
3	0.99366	0.99274	0.99159	54	0.90895	0.90488	0.90079
4	0.99197	0.99102	0.98984	55	0.90670	0.90262	0.89851
5	0.99033	0.98936	0.98815	56	0.90143	0.90034	0.89622
6	0.98877	0.98776	0.98651	57	0.90215	0.89805	0.89392
7	0.98726	0.98620	0.98491	58	0.89987	0.89576	0.89162
8	0.98581	0.98470	0.98336	59	0.89758	0.89346	0.88931
9	0.98442	0.98325	0.98185	60	0.89528	0.89115	0.88700
10	0.98307	0.98185	0.98038	61	0.89297	0.88883	0.88467
11	0.98176	0.98047	0.97893	62	0.89066	0.88651	0.88234
12	0.98049	0.97913	0.97752	63	0.88834	0.88418	0.88000
13	0.97925	0.97781	0.97612	64	0.88601	0.88185	0.87766
14	0.97803	0.97651	0.97474	65	0.88368	0.87950	0.87530
15	0.97683	0.97522	0.97336	66	0.88134	0.87716	0.87295
16	0.97563	0.97393	0.97199	67	0.87899	0.87480	0.87058
17	0.97444	0.97264	0.97061	68	0.87664	0.87244	0.86824
18	0.97324	0.97134	0.96922	69	0.87428	0.87008	0.86583
19	0.97203	0.97003	0.96782	70	0.87192	0.86770	0.86344
20	0.97080	0.96870	0.96640	71	0.86954	0.86532	0.86105
21	0.96956	0.96736	0.96497	72	0.86716	0.86292	0.85864
22	0.96829	0.96599	0.96352	73	0.86477	0.86052	0.85622
23	0.96699	0.96459	0.96203	74	0.86237	0.85812	0.85380
24	0.96566	0.96317	0.96052	75	0.85997	0.85570	0.85137
25	0.96430	0.96171	0.95897	76	0.85755	0.85328	0.84893
26	0.96289	0.96021	0.95739	77	0.85513	0.85084	0.84648
27	0.96145	0.95868	0.95577	78	0.85279	0.84840	0.84403
28	0.95997	0.95711	0.95412	79	0.85026	0.84595	0.84157
29	0.95845	0.95550	0.95244	80	0.84781	0.84349	0.83909
30	0.95688	0.95385	0.95071	81	0.84534	0.84101	0.83660
31	0.95526	0.95215	0.94894	82	0.84286	0.83852	0.83410
32	0.95360	0.95042	0.94713	83	0.84037	0.83602	0.83159
33	0.95191	0.94865	0.94529	84	0.83786	0.83350	0.82906
34	0.95017	0.94684	0.94342	85	0.83534	0.83097	0.82652
35	0.94839	0.94499	0.94152	86	0.83279	0.82842	0.82396
36	0.94657	0.94311	0.93957	87	0.83022	0.82583	0.82137
37	0.94471	0.94119	0.93760	88	0.82762	0.82323	0.81876
38	0.94282	0.93924	0.93560	89	0.82500	0.82060	0.81613
39	0.94089	0.93725	0.93356	90	0.82235	0.81795	0.81348
40	0.93893	0.93524	0.93151	91	0.81966	0.81527	0.81080
41	0.93694	0.93320	0.92943	92	0.81694	0.81255	0.80809
42	0.93491	0.93113	0.92732	93	0.81418	0.80979	0.80534
43	0.93286	0.92904	0.92519	94	0.81138	0.80700	0.80256
44	0.93078	0.92693	0.92305	95	0.80854	0.80417	0.79974
45	0.92868	0.92480	0.92088	96	0.80564	0.80129	0.79689
46	0.92655	0.92264	0.91870	97	0.80271	0.79838	0.79400
47	0.92441	0.92047	0.91650	98	0.79972	0.79541	0.79106
48	0.92225	0.91828	0.91429	99	0.79668	0.79240	0.78809
49	0.92006	0.91608	0.91207	100	0.79358	0.78933	0.78507
50	0.91787	0.91386	0.90983	...	......	......	......

* The table gives the specific gravity at the temperature indicated referred to water at 4°C. (U. S. Bureau of Standards.)

Table 2. Percentages by Volume at 60°F of Ethyl Alcohol Corresponding to Apparent Specific Gravities at Various Temperatures *

Apparent Specific Gravity	15.56/15.56	20/20	25/25	30/30	35/35	Apparent Specific Gravity	15.56/15.56	20/20	25/25	30/30	35/35
0.8040	97.85	97.12	96.31	95.50	94.69	0.7970	99.33	98.68	97.95	97.21	96.46
39	.87	.15	.33	.52	.72	69	.35	.70	.97	.23	.48
38	.89	.17	.36	.55	.74	68	.37	.72	.99	.25	.51
37	.91	.19	.38	.57	.77	67	.39	.75	98.02	.28	.53
36	.94	.22	.41	.60	.79	66	.42	.77	.04	.30	.56
35	.96	.24	.43	.62	.82	65	.44	.79	.06	.32	.58
34	.98	.26	.46	.65	.85	64	.46	.81	.08	.35	.60
33	98.00	.29	.48	.67	.87	63	.48	.83	.10	.37	.63
32	.03	.31	.50	.70	.90	62	.50	.85	.12	.39	.65
31	.05	.33	.53	.72	.92	61	.52	.87	.15	.42	.68
30	.07	.36	.55	.75	.95	60	.54	.89	.17	.44	.70
29	.09	.38	.58	.77	.98	59	.56	.91	.19	.46	.72
28	.11	.40	.60	.80	95.00	58	.58	.93	.21	.49	.75
27	.14	.43	.62	.82	.03	57	.60	.95	.23	.51	.77
26	.16	.45	.65	.85	.05	56	.61	.97	.26	.53	.80
25	.18	.47	.67	.87	.08	55	.63	99.00	.28	.56	.82
24	.20	.49	.70	.90	.11	54	.65	.02	.30	.58	.84
23	.22	.52	.72	.92	.13	53	.67	.04	.32	.60	.87
22	.25	.54	.74	.95	.16	52	.69	.06	.34	.62	.89
21	.27	.56	.77	.97	.18	51	.71	.08	.36	.65	.92
20	.29	.59	.79	96.00	.21	50	.73	.10	.39	.67	.94
19	.31	.61	.82	.02	.23	49	.75	.12	.41	.69	.96
18	.33	.63	.84	.05	.26	48	.77	.14	.43	.71	.99
17	.35	.66	.86	.07	.28	47	.79	.16	.45	.74	97.01
16	.38	.68	.89	.10	.31	46	.81	.18	.47	.76	.04
15	.40	.70	.91	.12	.34	45	.83	.20	.49	.78	.06
14	.42	.72	.94	.15	.36	44	.85	.22	.51	.80	.08
13	.44	.75	.96	.17	.39	43	.87	.24	.54	.83	.11
12	.46	.77	.98	.20	.41	42	.89	.26	.56	.85	.13
11	.48	.79	97.01	.22	.44	41	.91	.28	.58	.87	.15
10	.50	.81	.03	.25	.46	40	.93	.30	.60	.89	.18
09	.53	.84	.05	.27	.49	39	.95	.32	.62	.92	.20
08	.55	.86	.08	.29	.52	38	.97	.34	.64	.94	.22
07	.57	.88	.10	.32	.54	37	.99	.36	.66	.96	.25
06	.59	.90	.12	.34	.57	36	100.00	.38	.68	.98	.27
05	.61	.92	.15	.37	.59	35		.40	.70	98.01	.29
04	.63	.95	.17	.39	.62	34		.42	.73	.03	.32
03	.65	.97	.19	.42	.64	33		.44	.75	.05	.34
02	.67	.99	.22	.44	.67	32		.46	.77	.07	.36
01	.70	98.01	.24	.47	.69	31		.48	.79	.09	.39
00	.72	.03	.26	.49	.72	30		.50	.81	.12	.41
0.7999	.74	.06	.29	.51	.74	29		.52	.83	.14	.43
98	.76	.08	.31	.54	.77	28		.54	.85	.16	.46
97	.78	.10	.33	.56	.79	27		.56	.87	.18	.48
96	.80	.12	.36	.59	.82	26		.58	.89	.20	.50
95	.82	.14	.38	.61	.84	25		.60	.91	.23	.52
94	.84	.17	.40	.63	.87	24		.62	.93	.25	.55
93	.86	.19	.43	.66	.89	23		.64	.96	.27	.57
92	.88	.21	.45	.68	.92	22		.66	.98	.29	.59
91	.90	.23	.47	.71	.94	21		.68	99.00	.31	.62
90	.92	.26	.50	.73	.97	20		.70	.02	.33	.64
89	.95	.28	.52	.75	.99	19		.72	.04	.36	.66
88	.97	.30	.54	.78	96.02	18		.74	.06	.38	.68
87	.99	.32	.57	.80	.04	17		.76	.08	.40	.71
86	99.01	.34	.59	.83	.07	16		.78	.10	.42	.73
85	.03	.36	.61	.85	.09	15		.80	.12	.44	.75
84	.05	.39	.63	.87	.12	14		.82	.14	.46	.77
83	.07	.41	.66	.90	.14	13		.84	.16	.48	.80
82	.09	.43	.68	.92	.16	12		.86	.18	.51	.82
81	.11	.45	.70	.95	.19	11		.88	.20	.53	.84
80	.13	.47	.72	.97	.21	10		.90	.22	.55	.86
79	.15	.49	.75	97.00	.24	09		.92	.24	.57	.89
78	.17	.51	.77	.02	.26	08		.94	.27	.59	.91
77	.19	.54	.79	.04	.29	07		.96	.29	.61	.93
76	.21	.56	.81	.07	.31	06		.98	.31	.63	.95
75	.23	.58	.84	.09	.34	05		100.00	.33	.66	.98
74	.25	.60	.86	.11	.36	04			.35	.68	98.00
73	.27	.62	.88	.14	.38	03			.37	.70	.02
72	.29	.64	.90	.16	.41	02			.39	.72	.04
71	.31	.66	.93	.18	.43	01			.41	.74	.07
70	.33	.68	.95	.21	.46	00			.43	.76	.09

* Reprinted from the publication of the Association of the Official Agricultural Chemists.

TABLE 2 (*Continued*)

Apparent Specific Gravity	15.56/15.56	20/20	25/25	30/30	35/35	Apparent Specific Gravity	15.56/15.56	20/20	25/25	30/30	35/35
0.8180	94.43	93.58	92.64	91.71	90.79	0.8110	96.20	95.42	94.53	93.67	92.80
79	.46	.61	.67	.74	.82	09	.23	.44	.56	.69	.83
78	.48	.64	.70	.77	.85	08	.25	.47	.59	.72	.86
77	.51	.66	.72	.79	.88	07	.28	.49	.61	.75	.89
76	.53	.69	.75	.82	.91	06	.30	.52	.64	.77	.92
75	.56	.72	.78	.85	.94	05	.32	.54	.66	.80	.94
74	.59	.74	.81	.88	.97	04	.35	.57	.69	.83	.97
73	.61	.77	.84	.91	91.00	03	.37	.59	.72	.85	93.00
72	.64	.80	.86	.94	.03	02	.40	.62	.74	.88	.03
71	.66	.82	.89	.97	.06	01	.42	.64	.77	.91	.05
70	.69	.85	.92	92.00	.09	00	.45	.67	.79	.94	.08
69	.72	.88	.95	.03	.12	0.8099	.47	.69	.82	.96	.11
68	.74	.90	.97	.05	.14	98	.50	.72	.85	.99	.14
67	.77	.93	93.00	.08	.17	97	.52	.74	.87	94.02	.16
66	.79	.96	.03	.11	.20	96	.54	.77	.90	.04	.19
65	.82	.98	.06	.14	.23	95	.57	.79	.92	.07	.22
64	.84	94.01	.09	.17	.26	94	.59	.82	.95	.10	.25
63	.87	.04	.11	.20	.29	93	.61	.84	.98	.12	.27
62	.90	.06	.14	.22	.32	92	.64	.87	95.00	.15	.30
61	.92	.09	.17	.25	.35	91	.66	.89	.03	.17	.33
60	.95	.12	.20	.28	.38	90	.69	.92	.05	.20	.36
59	.97	.14	.22	.31	.40	89	.71	.94	.08	.23	.38
58	95.00	.17	.25	.34	.43	88	.73	.97	.10	.25	.41
57	.03	.20	.28	.36	.46	87	.76	.99	.13	.28	.44
56	.05	.22	.30	.39	.49	86	.78	96.02	.16	.31	.46
55	.08	.25	.33	.42	.52	85	.81	.04	.18	.33	.49
54	.10	.28	.36	.45	.55	84	.83	.07	.21	.36	.52
53	.13	.30	.39	.48	.58	83	.85	.09	.23	.39	.55
52	.15	.33	.41	.51	.61	82	.88	.11	.26	.41	.57
51	.18	.36	.44	.54	.64	81	.90	.14	.28	.44	.60
50	.20	.38	.47	.56	.66	80	.93	.16	.31	.47	.63
49	.23	.41	.50	.59	.69	79	.95	.19	.33	.49	.65
48	.25	.44	.52	.62	.72	78	.97	.21	.36	.52	.68
47	.28	.46	.55	.65	.75	77	97.00	.24	.39	.54	.71
46	.30	.49	.58	.68	.78	76	.02	.26	.41	.57	.73
45	.33	.51	.60	.70	.81	75	.04	.29	.44	.60	.76
44	.36	.54	.63	.73	.84	74	.07	.31	.46	.62	.79
43	.38	.57	.66	.76	.87	73	.09	.33	.49	.65	.81
42	.41	.59	.69	.79	.90	72	.11	.36	.51	.67	.84
41	.43	.62	.71	.82	.92	71	.14	.38	.54	.70	.87
40	.46	.64	.74	.84	.95	70	.16	.41	.56	.73	.90
39	.48	.67	.77	.87	.98	69	.18	.43	.59	.75	.92
38	.51	.70	.79	.90	92.01	68	.21	.46	.61	.78	.95
37	.53	.72	.82	.93	.04	67	.23	.48	.64	.80	.98
36	.56	.75	.85	.96	.07	66	.25	.50	.66	.83	94.00
35	.58	.77	.87	.98	.10	65	.28	.53	.69	.86	.03
34	.61	.80	.90	93.01	.13	64	.30	.55	.71	.88	.06
33	.63	.83	.93	.04	.15	63	.32	.58	.74	.91	.08
32	.66	.85	.95	.07	.18	62	.35	.60	.76	.93	.11
31	.68	.88	.98	.09	.21	61	.37	.63	.79	.96	.14
30	.71	.90	94.01	.12	.24	60	.39	.65	.81	.99	.16
29	.73	.93	.03	.15	.27	59	.42	.67	.84	95.01	.19
28	.76	.95	.06	.18	.30	58	.44	.70	.86	.04	.22
27	.78	.98	.09	.20	.33	57	.46	.72	.89	.06	.24
26	.81	95.01	.11	.23	.35	56	.49	.75	.91	.09	.27
25	.83	.03	.14	.26	.38	55	.51	.77	.94	.11	.30
24	.86	.06	.17	.29	.41	54	.53	.79	.96	.14	.32
23	.88	.08	.19	.31	.44	53	.55	.82	.99	.17	.35
22	.91	.11	.22	.34	.47	52	.58	.84	96.01	.19	.38
21	.93	.14	.24	.37	.50	51	.60	.86	.04	.22	.40
20	.96	.16	.27	.40	.53	50	.62	.89	.06	.24	.43
19	.98	.19	.30	.42	.55	49	.64	.91	.09	.27	.45
18	96.01	.21	.32	.45	.58	48	.67	.94	.11	.29	.48
17	.03	.24	.35	.48	.61	47	.69	.96	.14	.32	.51
16	.06	.26	.38	.51	.64	46	.71	.98	.16	.34	.53
15	.08	.29	.40	.53	.66	45	.73	97.01	.19	.37	.56
14	.10	.32	.43	.56	.69	44	.76	.03	.21	.40	.59
13	.13	.34	.46	.59	.72	43	.78	.05	.24	.42	.61
12	.15	.37	.48	.61	.75	42	.80	.08	.26	.45	.64
11	.18	.39	.51	.64	.78	41	.82	.10	.28	.47	.66
10	.20	.42	.53	.67	.80	40	.85	.12	.31	.50	.69

TABLE 3. Specific Gravity of Aqueous Solutions Referred to Water at the Same Temperature

Per Cent Alcohol	Sp. Gr. 15°/15° C	Sp. Gr. 20°/20° C	Sp. Gr. 25°/25° C	Per Cent Alcohol	Sp. Gr. 15°/15° C	Sp. Gr. 20°/20° C	Sp. Gr. 25°/25° C
0	1.00000	1.00000	1.00000	51	0.91635	0.91322	0.91026
1	0.99812	0.99813	0.99811	52	0.91412	0.91097	0.90799
2	0.99629	0.99629	0.99627	53	0.91189	0.90872	0.90571
3	0.99451	0.99451	0.99447	54	0.90964	0.90645	0.90343
4	0.99281	0.99279	0.99274	55	0.90738	0.90418	0.90113
5	0.99118	0.99113	0.99106	56	0.90512	0.90191	0.89883
6	0.98963	0.98955	0.98945	57	0.90285	0.89962	0.89654
7	0.98815	0.98802	0.98788	58	0.90058	0.89733	0.89423
8	0.98670	0.98653	0.98634	59	0.89830	0.89502	0.89191
9	0.98528	0.98505	0.98481	60	0.89601	0.89271	0.88959
10	0.98390	0.98361	0.98330	61	0.89371	0.89040	0.88725
11	0.98256	0.98221	0.98184	62	0.89139	0.88807	0.88491
12	0.98126	0.98084	0.98039	63	0.88907	0.88574	0.88256
13	0.97999	0.97948	0.97897	64	0.88674	0.88339	0.88020
14	0.97875	0.97816	0.97757	65	0.88441	0.88104	0.87783
15	0.97754	0.97687	0.97619	66	0.88207	0.87869	0.87547
16	0.97637	0.97560	0.97484	67	0.87971	0.87632	0.87309
17	0.97518	0.97431	0.97346	68	0.87736	0.87396	0.87071
18	0.97398	0.97301	0.97207	69	0.87500	0.87158	0.86833
19	0.97276	0.97169	0.97065	70	0.87263	0.86920	0.86593
20	0.97152	0.97036	0.96922	71	0.87025	0.86680	0.86352
21	0.97028	0.96901	0.96778	72	0.86785	0.86440	0.86110
22	0.96902	0.96763	0.96630	73	0.86545	0.86200	0.85869
23	0.96773	0.96624	0.96481	74	0.86304	0.85958	0.85626
24	0.96642	0.96483	0.96329	75	0.86063	0.85716	0.85383
25	0.96508	0.96339	0.96176	76	0.85822	0.85473	0.85140
26	0.96371	0.96190	0.96018	77	0.85579	0.85230	0.84895
27	0.96228	0.96037	0.95856	78	0.85336	0.84985	0.84650
28	0.96080	0.95880	0.95689	79	0.85092	0.84740	0.84404
29	0.95927	0.95717	0.95520	80	0.84846	0.84494	0.84157
30	0.95769	0.95551	0.95345	81	0.84599	0.84245	0.83909
31	0.95607	0.95381	0.95168	82	0.84350	0.83997	0.83659
32	0.95440	0.95207	0.94986	83	0.84101	0.83747	0.83408
33	0.95269	0.95028	0.94802	84	0.83850	0.83496	0.83156
34	0.95094	0.94847	0.94613	85	0.83598	0.83242	0.82902
35	0.94915	0.94662	0.94422	86	0.83343	0.82987	0.82646
36	0.94732	0.94473	0.94227	87	0.83086	0.82729	0.82389
37	0.94546	0.94281	0.94031	88	0.82826	0.82469	0.82128
38	0.94355	0.94086	0.93830	89	0.82564	0.82207	0.81865
39	0.94161	0.93886	0.93626	90	0.82299	0.81942	0.81600
40	0.93964	0.93684	0.93421	91	0.82030	0.81674	0.81331
41	0.93764	0.93479	0.93212	92	0.81759	0.81401	0.81060
42	0.93559	0.93272	0.93001	93	0.81484	0.81127	0.80785
43	0.93352	0.93062	0.92787	94	0.81205	0.80848	0.80507
44	0.93143	0.92849	0.92571	95	0.80922	0.80567	0.80225
45	0.92933	0.92636	0.92355	96	0.80636	0.80280	0.79939
46	0.92721	0.92421	0.92137	97	0.80344	0.79988	0.79648
47	0.92506	0.92204	0.91917	98	0.80045	0.79688	0.79349
48	0.92291	0.91986	0.91697	99	0.79739	0.79383	0.79045
49	0.92075	0.91766	0.91475	100	0.79429	0.79074	0.78736
50	0.91856	0.91546	0.91251	...	......	......	......

TABLE 4. Specific Gravity of Mixtures of Methyl Alcohol and Water by Volume and by Weight *

Per Cent Alcohol by Weight	Per Cent Alcohol by Volume	Specific Gravity	Per Cent Alcohol by Weight	Per Cent Alcohol by Volume	Specific Gravity
1	1.25	0.99727	51	58.74	9.91653
2	2.50	0.99543	52	59.76	0.91451
3	3.75	0.99370	53	60.77	0.91248
4	4.99	0.99198	54	61.78	0.91044
5	6.22	0.99029	55	62.78	0.90839
6	7.45	0.98864	56	63.78	0.90631
7	8.68	0.98701	57	64.77	0.90421
8	9.91	0.98547	58	65.75	0.90210
9	11.13	0.98394	59	66.73	0.89996
10	12.35	0.98241	60	67.69	0.89781
11	13.56	0.98093	61	68.65	0.89563
12	14.77	0.97945	62	69.61	0.89341
13	15.98	0.97802	63	70.55	0.89117
14	17.18	0.97660	64	71.49	0.88890
15	18.38	0.97518	65	72.42	0.88662
16	19.58	0.97377	66	73.34	0.88433
17	20.77	0.97237	67	74.26	0.88203
18	21.96	0.97096	68	75.17	0.87971
19	23.15	0.96955	69	76.08	0.87739
20	24.33	0.96814	70	76.98	0.87507
21	25.51	0.96673	71	77.86	0.87271
22	26.69	0.96533	72	78.75	0.87033
23	27.86	0.96392	73	79.62	0.86792
24	29.03	0.96251	74	80.48	0.86546
25	30.19	0.96108	75	81.34	0.86300
26	31.35	0.95963	76	82.18	0.86051
27	32.51	0.95817	77	83.02	0.85801
28	33.66	0.95668	78	83.86	0.85551
29	34.81	0.95518	79	84.68	0.85300
30	35.95	0.95366	80	85.50	0.85048
31	37.09	0.95213	81	86.31	0.84794
32	38.22	0.95056	82	87.11	0.84536
33	39.35	0.94896	83	87.90	0.84274
34	40.48	0.94734	84	88.68	0.84009
35	41.59	0.94570	85	89.45	0.83742
36	42.71	0.94404	86	90.21	0.83475
37	43.82	0.94237	87	90.97	0.83207
38	44.92	0.94067	88	91.72	0.82937
39	46.02	0.93894	89	92.46	0.82667
40	47.11	0.93720	90	93.19	0.82396
41	48.20	0.93543	91	93.92	0.82124
42	49.28	0.93365	92	94.63	0.81849
43	50.35	0.93185	93	95.33	0.81568
44	51.42	0.93001	94	96.02	0.81285
45	52.49	0.92815	95	96.70	0.80999
46	53.54	0.92627	96	97.37	0.80713
47	54.60	0.92436	97	98.04	0.80428
48	55.64	0.92242	98	98.70	0.80143
49	56.68	0.92048	99	99.35	0.79859
50	57.71	0.91852	100	100.00	0.79577

* Giving the specific gravity at 15°C referred to water at 4°C. (Calculated from values by Doroshevski and Rozhdestvenski, Dittmar and Fawsitt.)

TEST FOR MIXED ALCOHOLS (HYDROXYL NUMBER)

Method L.I.C. No. 240.01C is intended primarily for the examination of specification alcohols of low molecular weight containing not over 5 per cent water. This method is essentially the same, but contains added directions so that the test may be applied to any type of primary or secondary alcohol in the presence of any amount of water. The test is based on the fact that the presence of alcohol will reduce the acidity of acetic anhydride, as shown in the following equations:

$$(CH_3CO)_2O + H_2O \longrightarrow 2\,CH_3COOH$$

$$(CH_3CO)_2O + ROH \longrightarrow CH_3COOR + CH_3COOH$$

The decrease in acidity of a given amount of acetic anhydride is thus a measure of the alcohol content of the sample under test.

This procedure is applicable to any type of primary or secondary alcohol containing only one hydroxyl group. It is also suitable for most phenolic hydroxyl compounds. Hydrocarbons, acetals, and esters do not interfere. A correction must be made for natural acidity or alkalinity. Mercaptans, amines, some ethers, and aldehydes (in concentrations higher than 10 per cent) interfere. Tertiary alcohols are incompletely determined.

Definition of Terms

When the sample under test consists of a mixture of alcohols, or when the type of alcohol in the sample is unknown, it is customary to report the result in either of the following terms:

Hydroxyl Number—It will be noted from the equation that the presence of one molecule of alcohol in the sample will reduce the acidity of the titration mixture by an amount equivalent to one molecule of acetic acid, regardless of the nature of the alcohol. This loss in acidity, expressed as the number of milligrams of potassium hydroxide required to neutralize the corresponding amount of acetic acid, is known as the *hydroxyl number,* and is usually reported as milligrams of KOH per gram of sample.

Centiequivalents—Most volumetric work is based on the familiar equation: *milliliters times normality equals number of milliequivalents.* In turn, the number of milliequivalents in 1 g of a sample, multiplied by the equivalent weight will give the weight in milligrams of the compound under test in 1 g of sample. By expressing this in centiequivalents instead of milliequivalents, the final result will be obtained in centigrams per gram, which is the same as per cent by weight.

Apparatus

(1) *Weighing Bottle*—A glass-stoppered bottle should be used.

(2) *Burette*—For the caustic solution, some type of automatic burette should be available, with a bulb of at least 50-ml capacity. When not in use, precautions should be taken to prevent the stopcock from sticking.

(3) *Reflux Assembly*—This consists of a 250-ml Erlenmeyer flask fitted to a 20-in. Allihn condenser by a ground-glass joint. This assembly is available at the Corning Glass Works (the flask under item 5000, and the condenser under 2440).

Reagents

(1) *Acetylating Reagent*—Shake 2 liters of reagent-grade pyridine with a few grams of solid caustic soda and filter the solution through a double thickness of dry filter paper. Mix 880 ml of the filtrate with 120 ml of chemically pure acetic anyhydride and store the resulting solution in an amber-colored bottle.

(2) *Standard Sodium Hydroxide Solution*—Prepare and standardize as directed in L.I.C. No. 240.01.

(3) *Indicator*—Dissolve 0.1 g of phenolphthalein in 100 ml of 91 per cent neutralized isopropyl alcohol.

For best results, samples of ethyl and/or methyl alcohol should not contain over 5 per cent water. As the molecular weight of the sample increases the percentage of water may be raised proportionately, but in no case should the water content be higher than 20 per cent. The water content, if unknown, may be determined as outlined in L.I.C. No. 150.04A.

In most cases, dilute solutions of alcohol can be concentrated by *salting out* with potassium carbonate. To do this, measure an appropriate amount of the sample into a glass-stoppered graduate, note the volume carefully, and slowly add the solid reagent until there is an undissolved excess on the bottom. Stopper the cylinder, shake, and allow to settle. The liquid will separate into two layers, with the upper one containing the alcohol. Note the volume of the upper layer, and then analyze it for its alcohol content. Determine and record the specific gravity of both the sample as received and the separated alcohol layer. When isopropyl alcohol is salted out, the lower layer at 35°C will contain 0.23 per cent alcohol; for accurate results this alcohol should be included in the total. The higher alcohols separate with only negligible loss.

This procedure cannot be applied to methyl and ethyl alcohols. Methyl alcohol will not *salt out* at all and ethyl alcohol will contain about 9 per cent water. These two should be concentrated by distillation in an efficient column.

Procedure

(1) Determine the proper size sample from the chart of Figure 1. If the alcohol content is entirely unknown, a trial test may be necessary. Weigh the proper amount into each of two tared weighing bottles.

In the case of methyl alcohol, the use of sealed glass ampoules, instead of weighing bottles, is suggested.

(2) Measure accurately 25 ml of the acetylating reagent into each of three Erlenmeyer flasks. Reserve one of these as a blank and place weighed samples into the other two. Put two drops of water into the flask reserved for the blank and connect all three to condensers. Lubricate the ground-glass joints with a few milliliters of chemically pure pyridine.

The water is added to the blank to prevent discoloration of the mixed liquids and also to inhibit a reaction between the dry pyridine and the acetic acid.

(3) Reflux the samples gently for at least 30 minutes, preferably 1 hour. During this time, weigh out another portion of the sample, heat to boiling, allow to cool, add a few drops of the indicator, and note whether the sample is acid or alkaline. If

the sample is either acid or alkaline, titrate with standard alkaline or acid solution, as necessary, and record the total volume of standard solution required.

If the analysis is made on a part of the sample as received, another portion should be taken for the determination of free acidity. If the sample under test is a part of a concentrate obtained either by "salting out" or by distillation, the free-acidity measurement should be made on another portion of the concentrate. In most cases, this correction will be low, so that it is customary to use an extremely dilute standard solution and to take a large weight of sample. If the correction is high, the operator may take a smaller weight of sample, or use a stronger standard solution, or both.

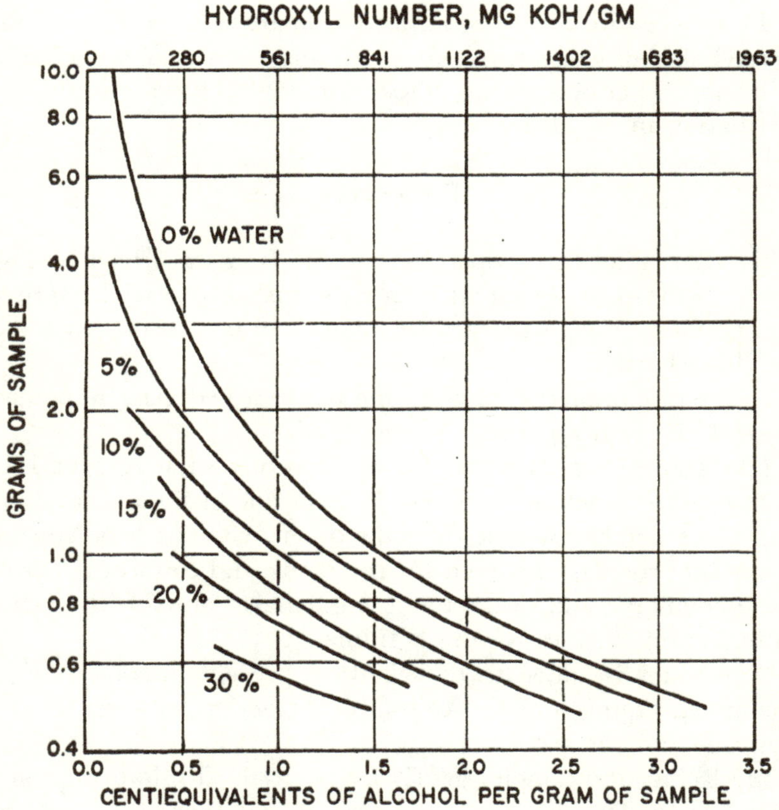

FIGURE 1. Conversion Chart—Hydroxyl Number to Centiequivalents

(4) At the end of the refluxing period, rinse each condenser and joint with 50 to 100 ml of distilled water, adding the rinsings to the proper flask. Allow to cool. When the sample contains suspended oily matter, add 25 ml of petroleum ether to avoid a holdup of acetic acid. Titrate the contents of each flask to a faint-pink end point with the standard caustic solution, using phenolphthalein as the indicator. Note and record the necessary volume of alkali for each flask.

Samples that are dark colored may be titrated potentiometrically to an end point of *p*H 9.

Calculations

(1) Determine the amount of free acidity or alkalinity by the equation:

$$F = \frac{A \times N \times 0.1}{S}$$

where

- A is the volume of standard acid or alkali, as necessary.
- N is the normality of this solution.
- S is the weight of the sample in grams for this determination.
- F is the factor to be added or subtracted to correct for any free acidity or alkalinity. (The value of 0.1 is included in the numerator to change milliequivalents to centiequivalents. F is to be subtracted when the test sample is alkaline, and added when it is acid.)

(2) Calculate the centiequivalents in the sample by the equation:

$$H = \frac{(B - S) \times N \times 0.1}{W} \pm F$$

where

- B is the volume in milliliters of standard caustic required by the blank.
- S is the corresponding volume for the sample.
- W is the weight in grams of the sample for this determination.

H is the number of centiequivalents of the hydroxyl ion present in 1 g of the sample.

When the sample under test is a concentrate obtained either by distillation or *salting out,* calculate the value for H to that of the original sample by:

$$H' = \frac{(\text{Volume in milliliters of the dehydrated layer}) \times (\text{sp. gr.*}) \times H}{(\text{Volume in milliliters of original sample}) \times (\text{sp. gr.*})}$$

(3) When the sample under test is a mixture of alcohols, multiply the value H (or H′, as applicable) by 561 to determine the *hydroxyl number.*

When the sample consists of one alcohol of known composition, multiply the value H (or H′) by the proper molecular weight to determine the weight percentage of the alcohol in the sample. The molecular weights of some of the common alcohols are listed in Table 1.

TABLE 1

C_1 (Methyl)	32.03	C_5	88.09	C_9	144.55
C_2 (Ethyl)	46.05	C_6	102.21	C_{10}	158.67
C_3 (Propyl)	60.06	C_7	116.32	C_{11}	172.78
C_4 (Butyl)	74.08	C_8	130.44	C_{12}	186.90

The calculations may, at first glance, appear confusing, so that the following explanation may be helpful. Let us assume that the sample under test is absolutely neutral, that 85 ml of caustic was required for the blank, and that 80 ml was needed for the sample. The acidity due to the alcohol would, therefore, be 85 − 80, or 5 ml of 0.8 N sodium hydroxide solution. However, samples are seldom absolutely neutral; in most cases, there is a slight amount of natural acidity. The final titration of the sample would, therefore, include both the natural acidity and the acidity formed by the reaction, so that the volume for the sample might be, say, 80.3. In this case, the difference between the titrations for the blank and the sample would be low in an amount corresponding to the natural acidity of the sample as received. In the given example, the final titration would be 4.7 ml, instead of 5.0. Therefore, when the sample is naturally acid, a correction

* Specific gravity.

must be added to the observed difference between the titrations for the blank and the sample. Of course, when the sample is alkaline, the reverse is true; the volume of standard alkali needed for the sample will be less, the difference will be larger, and the correction must be subtracted.

In specification alcohols, the amount of natural acidity or alkalinity is usually quite small. If this determination were made in the normal fashion (1 g of sample and using 0.8 N caustic), the error in the titration would be greater than the size of the correction. It is, therefore, customary to make this determination on a much larger sample, using a weaker solution. For mixed or non-specification alcohols, the amount of acidity may be much higher. The titrating conditions (size of sample, strength of solution, etc.) are, therefore, left to the judgment of the operator. Since the titrating conditions are different, the observed volume of standard solution required for the correction cannot be applied directly to the volume for the sample. It is, therefore, necessary to determine the values for the sample and the correction in common terms (centiequivalents) and then to add or subtract as necessary to obtain the net centiequivalents for the test sample.

ACETONE IN ISOPROPYL ALCOHOL BY ULTRAVIOLET SPECTROMETRY

This procedure is used to measure small amounts of acetone in refined isopropyl alcohol.

It follows the familiar base-line technique. Acetone will absorb sharply over a broad range, with a peak setting at or near 275 mμ. The absorbancy at this setting, as compared with the base line drawn from 240 to 320 mμ, is proportional to the concentration of acetone in the sample.

Other free saturated aliphatic carbonyl compounds will appear as acetone. Latent carbonyl bodies, such as acetals and ketals, are not determined by the method.

Apparatus

(1) *Ultraviolet Spectrophotometer*—Any convenient type may be used. The Model DU, available from the Beckman Instruments Co., So. Pasadena, Calif., is satisfactory.

Reagents

(1) *Acetone,* ACS reagent-grade.

(2) *Isopropyl Alcohol*—Spectrophotometric grade is available from Merck.

Calibration

(1) Pour about 10 ml of spectrophotometric-quality isopropyl alcohol into a 100-ml volumetric flask, insert the stopper, and weigh the flask to the nearest milligram. Add some 0.4 g of acetone, and weigh again. Record the difference as the weight of acetone added. Fill the flask to the mark with more alcohol and mix thoroughly.

(2) Using this solution and more alcohol, prepare a set of standards covering the range of 4.0 to 0.1 g of acetone per liter.

(3) Fill a 1-cm cell with the strongest of the solutions and scan

in the spectrophotometer in the range of 235 to 325 mμ. Use the pure alcohol as the reference. Note and record particularly the exact absorbancies at or near 240 and 320, and the peak height at or near 275.

(4) Repeat this procedure with each of the remaining solutions. Subtract the absorbancy at a base line, drawn at 240 to 320, from that of the solution at 275, and record the net absorbancy of each solution.

(5) Prepare a graph plotting net absorbancy against grams of acetone per liter of solution.

Procedure

(1) Fill a cell with the sample under test, and note the absorbancy at the critical settings, using air (no empty cell in position) as reference.

(2) Subtract the base-line correction from the absorbancy observed at 275, and record the difference. Find the acetone content from the graph, in grams per liter.

If the observed content is beyond the range of the graph, dilute the sample with a measured amount of alcohol and note the dilution factor. Fill the cell with this solution and a similar cell with the alcohol used. Put the second into the reference spot and examine as usual. Make the proper correction in the final calculations.

(3) Determine the density of the sample at test temperature, using either a Westphal balance or a pycnometer.

(4) Calculate the acetone content of the sample by the equation:

$$A = \frac{G \times 100}{D \times 1000}$$

where

G is the acetone content, in grams per liter, as found from the graph.

D is the density of the sample at test temperature.

A is the acetone content expressed as weight per cent.

Duplicate results should not differ from one another by more than 5 per cent of the observed content. The accuracy of the test is in the same order of magnitude.

ESTER CONTENT OF SOLVENTS

Acetic esters of the lower aliphatic alcohols are determined by alkali saponification. This method is based on ASTM D-268. It is suitable for determining the secondary alkyl acetate content of solvents or solvent blends. The presence of free acid or alkali in the sample before the addition of the KOH will introduce an error in the analysis. Therefore, any free acid or alkali in the original sample should be quantitatively determined and the result used to correct the final calculation when the percentage by weight of acetic ester is determined.

Apparatus

(1) *Weighing Bulbs or Ampoules*—These are made of thin-walled, Pyrex or alkali-free soft glass and blown to a diameter of about $\frac{3}{4}$ in. with a stem of about 3 in. length. Ampoules of this size will handle up to 2 g samples.

(2) *Refluxing Equipment*—This consists of a 500-ml Erlenmeyer flask fitted to an Allihn condenser of 24 in. length. The flask and condenser are joined together with a standard-taper ground-glass joint.

Reagents

(1) *Standard Potassium Hydroxide Solution*—A 0.5 N aqueous potassium hydroxide solution is prepared by adding 454 g (1 lb) of reagent-grade potassium hydroxide pellets to 450 ml of water in a 1-liter flask. The flask is closed with a rubber stopper fitted with a soda-lime tube and allowed to stand for several days. A sirupy solution will form, from which practically all carbonate will precipitate. When no more precipitation takes place, 28 ml of the supernatant sirupy solution is taken for each liter of 0.5 N potassium hydroxide solution desired and diluted to the predetermined volume with distilled water which has been freshly boiled and cooled. This solution is kept in a large bottle or carboy and is protected by a soda-lime tube.

Potassium acid phthalate or benzoic acid is used as a primary standard in standardizing the potassium hydroxide solution. If the phthalate, which is the preferred standard, is used, it is dried to constant weight in an oven at 105°C. Two or three portions, 4 g each, are weighed to ±0.0001 g in weighing bottles and transferred quantitatively to separate 250-ml Erlenmeyer flasks. About 100 ml of carbonate-free distilled water is poured into each flask and the salt dissolved. Four drops of the phenolphthalein are added into each flask. The acid solution is constantly agitated in the flask, heated, and titrated while hot with the potassium hydroxide solution. When the end point is approached, the flask is washed down with distilled water and the titration continued to the end point. The normality of the base is calculated from the following equation:

$$\text{Normality of KOH} = \frac{\text{weight of } KHC_8H_5O_4}{0.20422 \times \text{milliliters of KOH solution}}$$

If benzoic acid is used as a primary standard, two or three portions, 2.5 g each, are weighed as before and dissolved in 25 ml of carefully neutralized isopropyl alcohol. Four drops of the phenolphthalein indicator and about 75 ml of carbonate-free distilled water are added and the solution titrated just below the boiling point with KOH. The normality of the base is calculated from the following equation:

$$\text{Normality of KOH} = \frac{\text{weight of } H{\cdot}C_7H_5O_2}{0.12205 \times \text{milliliters of KOH}}$$

(2) *Standard Hydrochloric Acid Solution*—A 0.5 N HCl solution is prepared by adding about 38 ml of concentrated chemically pure HCl to 400 ml of distilled water for each liter of 0.5 N HCl desired. The solution is diluted with distilled water to the predetermined volume and then standardized by titrating with a standard NaOH or KOH solution which has been checked against a primary standard, such as potassium acid phthalate or benzoic acid within 10 days.

(3) *Alcohol*—The alcohol may be absolute methyl, 95 per cent ethyl, or 99 per cent isopropyl alcohol. It is neutralized by adding sufficient alkali to give a very faint pink color with phenolphthalein.

(4) *Phenolphthalein*—An indicator solution is prepared by dissolving 0.1 g of phenolphthalein in 100 ml of 91 per cent refined isopropyl alcohol. This solution is neutralized by adding sufficient alkali to give a very faint pink color.

Procedure

(1) Three weighing bulbs or ampoules of approximately the same size are selected. Two are weighed accurately to the nearest 0.1 mg and the third stored in a desiccator.

(2) The two ampoules which have been weighed are partially filled, separately, with 1 to 2 g of the sample, being careful to avoid a yellow sodium flame when the bulb is heated. A small air space should be left in the glass bulb, after which the stem is slightly warmed to expel any liquid.

(3) The tip of the stem should be sealed in a flame, again avoiding the yellow flame. The sealed tip is allowed to cool before the liquid in the bulb is allowed to come in contact with it. The two ampoules are wiped dry with a clean cloth and reweighed to the nearest 0.1 mg. The weights of the two samples are calculated and recorded.

(4) Three units of the refluxing equipment, as described under APPARATUS (2), are prepared and one of the three ampoules is placed in each of the three flasks. By means of an accurate burette, at least one and one half times the theoretical quantity of standard KOH solution required for complete saponification is placed into each flask. The amount can be estimated from the weight of the sample and Table 1 which lists equivalents based on pure acetates.

TABLE 1

1 g of pure ethyl acetate	would require 22.8 ml 0.5 N KOH
1 g of pure isopropyl acetate	would require 19.6 ml 0.5 N KOH
1 g of pure secondary butyl acetate	would require 17.2 ml 0.5 N KOH
1 g of pure secondary amyl acetate	would require 15.4 ml 0.5 N KOH
1 g of pure secondary hexyl acetate	would require 13.9 ml 0.5 N KOH
1 g of pure secondary heptyl acetate	would require 12.6 ml 0.5 N KOH

The neutralized alcohol will be added to each flask (including the blank) in an amount sufficient to effect complete solution of the sample when the ampoules are broken.

(5) The ampoule in each flask is crushed by means of a separate stirring rod being careful to break up the entire length of the stem. Each rod is carefully rinsed with about 5 ml of distilled water.

(6) Then each flask is connected to its condenser and the units set up on a hot plate or steam bath and refluxed for 1 to 4 hours, depending on the ester being tested. During the refluxing period, just enough heat is applied to prevent any loss of volatile matter at the top of each condenser.

(7) At the end of the refluxing period, the set-ups are allowed to cool to room temperature, each condenser rinsed with distilled water, the flasks disconnected, and 3 or 4 drops of the phenolphthalein added to the contents of each flask.

(8) The excess KOH in each flask is determined by titration with the standard HCl solution. The number of milliliters required for each titration is recorded.

(9) The results are calculated and reported on the basis of the following equations:

$$\text{Milliliters of HCl for sample} \times \text{N. of HCl} = A$$

$$\text{Milliliters of HCl for blank} \times \text{N. of HCl} = B$$

Then: B = A = milliliters of N/1 KOH used by the Sample = C

And:

$$\frac{C \times 0.06003 \times 100}{\text{wt. of sample in grams}} = \text{Per cent by Weight Acetic Acid}$$

To convert the acetic acid percentage into the corresponding alkyl ester percentage, it is necessary to multiply by the following factors:

% Ethyl Acetate by Weight = % Acetic Acid × 1.470
% Isopropyl Acetate by Weight = % Acetic Acid × 1.700
% Secondary Butyl Acetate by Weight = % Acetic Acid × 1.935
% Secondary Amyl Acetate by Weight = % Acetic Acid × 2.170
% Secondary Hexyl Acetate by Weight = % Acetic Acid × 2.400
% Secondary Heptyl Acetate by Weight = % Acetic Acid × 2.630

TESTS FOR SPECIFICATION GRADES OF ETHYL ETHER

Several methods for the examination of specification-grade samples of ethyl ether have been combined into one unit.

Acetylene

Add 10 drops of freshly prepared 0.1 per cent methyl red solution into each of two 250-ml Erlenmeyer flasks containing 50 ml of 50 per cent alcohol (SDA 2B). Reserve one flask as a blank and place into the other 10.0 ml of the sample under test. Neutralize any acidity and match the colors of the liquids in the two flasks exactly by adding 0.02 N sodium hydroxide dropwise from a burette, disregarding the amount put into either flask. Pour 10.0 ml of about 0.06 N alcoholic silver nitrate solution into each flask and titrate both to the same yellow end point with the 0.02 N NaOH, recording the amount for each. Calculate the acetylene content by the equation:

$$A = \frac{(V - v) \times N \times 0.0130 \times 100}{S \times G}$$

where

- V is the volume in milliliters of the NaOH required for the sample.
- v is the corresponding volume for the blank.
- N is the exact normality of the NaOH solution.
- S is the volume of sample, in milliliters.
- G is the specific gravity of the sample, as determined on a separate portion.
- A is the acetylene content of the sample, expressed as weight per cent.

Acidity

Prepare the indicator by dissolving 0.1 g of bromothymol blue in 20.0 ml of hot alcohol (SDA 2B) and adding distilled water to make 100.0 ml of solution. The distilled water should be free of carbon dioxide, and the solution should be stored in a tightly stoppered bottle.

Into a colorless glass flask or cylinder which can be tightly stoppered, put 2 drops of the indicator and 10 ml of distilled water. Add, dropwise, 0.01 N sodium hydroxide solution until the first permanent blue color is developed, disregarding the amount actually used. Add 50 ml of the sample, stopper the flask, and shake thoroughly. Titrate with the 0.01 N NaOH to the first blue color which persists for several minutes. Calculate the acidity of the sample by the equation:

$$A = \frac{V \times N \times 0.060 \times 100}{S \times G}$$

where

- V is the volume in milliliters of NaOH used for the titration of the sample.
- N is the exact normality of the NaOH solution.
- S is the volume of the sample, in milliliters.
- G is the specific gravity of the sample, as determined on a separate portion.
- A is the acidity, expressed as weight per cent of acetic acid.

Appearance

Fill a 25-ml test tube made of colorless glass with the sample, and another with distilled water. Compare the liquids in the two tubes. The sample should not be darker in color than the water and should not contain visible suspended matter.

Chlorides

Put 10 ml of the sample into a test tube and add 4 to 5 ml of a 10 per cent aqueous solution of silver nitrate. If an opalescence appears, add 1 drop of nitric acid and stir with a glass rod. A persistent opalescence indicates the presence of chlorides. (A temporary opalescence may appear when the sample is slightly alkaline.)

Heavy Hydrocarbons

Put 150 ml of the sample and 75 ml of distilled water into a 300-ml distillation flask and assemble the distillation apparatus. This assembly should include a Vigreaux column (18 by ½ in.), a distillation head which can be cooled with dry ice, a thermometer, and provision for controlling the reflux ratio. Boil the sample for 10 minutes without collecting any distillate and then adjust the reflux ratio at 5 to 1. Distill under these conditions until the temperature of the overhead reaches 75°C. Change the receiver and continue the distillation, collecting the liquid in a 25-ml graduate, until no more hydrocarbons appear in the distillate. Note and record the volume of the hydrocarbons. Calculate the hydrocarbon content of the sample by the equation:

$$H = \frac{V \times d \times 100}{S \times D}$$

where

- V is the volume of hydrocarbons, in milliliters.
- d is the density of the hydrocarbons determined on a Davidson gravitometer. This determination need not be made on every run; a typical value is adequate.
- S is the volume of sample, in milliliters.
- D is the density of the sample.
- H is the hydrocarbon content expressed as weight per cent.

Nonvolatile Matter

Put 100 ml of the sample into a tared 100-ml evaporating dish and place the dish on a steam bath. Evaporate the ether, dry the dish for 1 hour in an oven at 100°C, cool in a desiccator, and reweigh. Determine the weight of 100 ml of the sample from the specific gravity and calculate the weight percentage of nonvolatile matter in the sample.

Peroxides

Put 10 ml of the sample into a 10-ml glass-stoppered graduate and add 1 ml of a solution prepared by dissolving 0.5 g of potassium iodide and 0.5 g of cadmium iodide in 9.0 ml of distilled water. Put the cylinder into a dark place and leave it there for 1 hour, shaking occasionally. If neither layer darkens, peroxides are absent.

Specific Gravity

Determine the specific gravity by means of a Westphal balance, as described under *240.03A*. Unless otherwise specified, test conditions are 20°/20°C.

DETERMINATION OF THE CARBONYL COMPOUNDS IN PETROCHEMICAL PRODUCTS (CARBONYL NUMBER)

In the production of certain chemicals and solvents from petroleum, some of the possible impurities are the types of compounds containing a carbonyl (CO) group. All of these will react with hydroxylamine hydrochloride according to the following equations:

Aldehydes—

$$RCHO + H_2NOH{\cdot}HCl \longrightarrow RCH{=}NOH + H_2O + HCl$$

Ketones—

$$R_2CO + H_2NOH{\cdot}HCl \longrightarrow R_2C{=}NOH + H_2O + HCl$$

Acetals—

$$RCH(OR)_2 + H_2O \longrightarrow RCHO + 2ROH$$

In the last case, one molecule of an acetal will react with one of water in an acid solution to form one molecule of an aldehyde, which, in turn, will react according to the first equation. Since carbonyl compounds can be present in specification-grade chemicals in only minute amounts, it is customary to determine all types together and to report the sum of all three, with no further separation. For the examination of relatively pure ketones, refer to L.I.C. No. 230.01A.

The final result may be reported in either of the following terms:

Centiequivalents—Most volumetric analyses are based on the familiar equation: *milliliters times normality equals number of milliequivalents.* In turn, the number of milliequivalents of any compound in 1 gram of sample, multiplied by the equivalent weight, will give the weight in milligrams of the compound in question. By using centiequivalents instead of milliequivalents, the final result will be obtained in centigrams per gram, which is the same as per cent by weight.

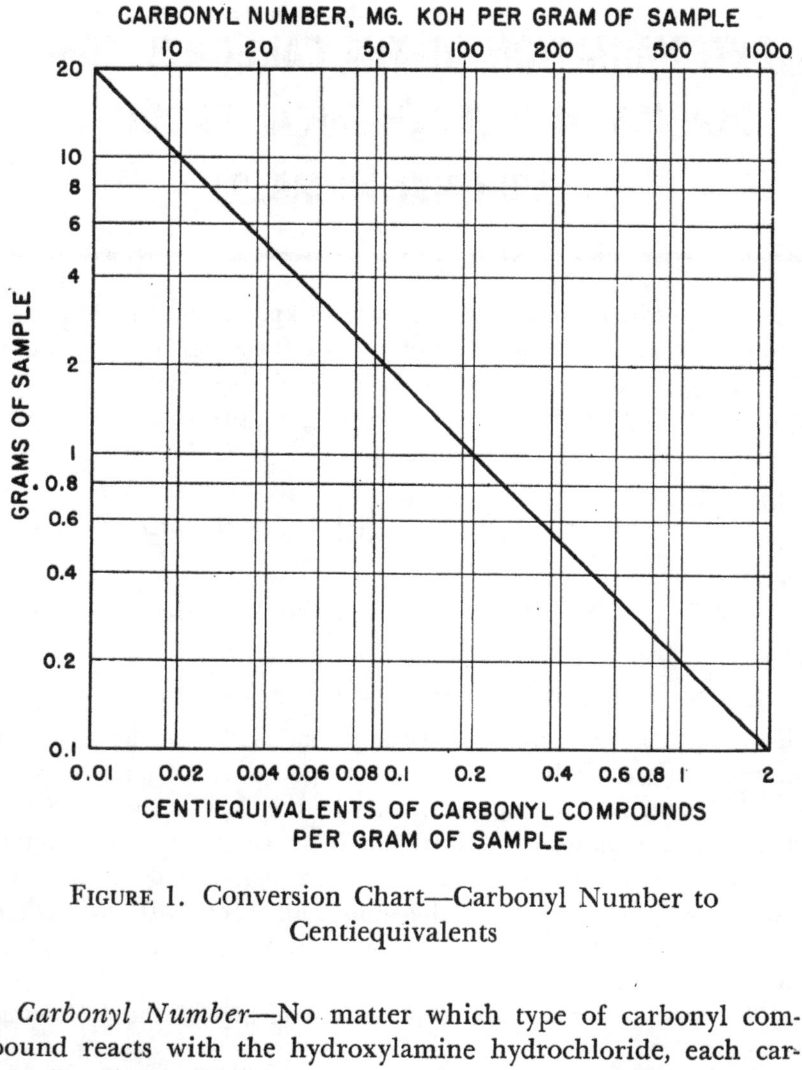

FIGURE 1. Conversion Chart—Carbonyl Number to Centiequivalents

Carbonyl Number—No matter which type of carbonyl compound reacts with the hydroxylamine hydrochloride, each carbonyl radical will liberate one molecule of hydrogen chloride. The amount of potassium hydroxide necessary to neutralize this acid is, therefore, a measure of the total carbonyl content of the sample. This value, expressed in milligrams of KOH per gram of sample, is known as the *carbonyl number.* It is usually reported as the total acid derived from all three types of compounds, with no distinction as to the amount produced by each.

Apparatus

(1) *Sample Container*—A weighing bottle fitted with a ground-glass stopper is usually satisfactory. For very volatile solvents, a sealed glass ampoule may be used.

(2) *Erlenmeyer Flasks*—Glass-stoppered, of 250-ml capacity.

Reagents

(1) *Hydroxylamine Hydrochloride*—Available from Baker Chemical Company. Dissolve 20 g of the solid in 1 liter of distilled water.

(2) *Sodium Hydroxide, 0.1 N Solution*—Prepare and standardize as directed in L.I.C. No. 230.01, taking 5.2 ml of the concentrated solution for each liter of 0.1 N solution desired.

(3) *Hydrochloric Acid, 0.1 N*—Mix 9 ml of the concentrated acid with sufficient distilled water to make 1 liter and standardize against the standard sodium hydroxide.

(4) *Methyl Orange–Xylene Cyanole Indicator*—Available from Eastman in the form of a concentrated solution; mix this solution with distilled water in the ratio of one to five.

If preferred, the solution may be prepared by dissolving 0.10 g of methyl orange and 0.14 g of xylene cyanole FF in a small amount of pure alcohol and diluting to 1 liter with distilled water.

Procedure

(1) Determine the proper size of sample from the graph of Figure 1.

(2) Measure exactly 25 ml of the hydroxylamine hydrochloride solution into a 250-ml flask, and add 2 drops of the indicator and the proper amount of the test sample.

In most cases, a tared weighing bottle may be used to measure the sample to be tested. For very volatile samples with a high carbonyl number, a sealed glass ampoule should be used, while nonvolatile samples of low carbonyl number may be measured into the flask by a pipette or graduate.

(3) Without delay, add 25 ml of carbonyl-free ethyl alcohol and put the flask under a reflux assembly. Heat at reflux temperature for 5 hours.

(4) During this time, weigh approximately the same amount of the sample into a separate clean container, add a few drops of the indicator, and note whether or not there is any change of color. If there is any indication of free acidity or alkalinity, titrate the sample with standard alkali or acid solution, as necessary, and note the volume of solution required.

(5) At the end of 5 hours of refluxing, remove the flask from the hot plate, add exactly 50 ml of 0.1 N NaOH solution, and allow to cool with occasional shaking.

After the addition of the alkali, the indicator color should be green. If pink, the sample size is too large; discard and repeat with a smaller sample.

(6) When the liquid in the flask is cool, titrate with standard 0.1 N acid to a slate-blue end point. Note and record the necessary volume of acid for the titration.

(7) Make a blank test on the same volumes of all the reagents, but without any sample. Record the volume of 0.1 N acid required for the blank.

If preferred, the titration may be done potentiometrically to a *p*H of 3.5.

Calculations

(1) Determine the amount of free acid or alkali by the equation:

$$F = \frac{A \times N \times 0.1}{S}$$

where

- A is the volume in milliliters of either standard alkali or acid.
- N is the normality of this solution.
- S is the weight of this sample in grams.
- F is the factor to correct for the free acidity or alkalinity. The value for F is expressed in centiequivalents per gram and is to be subtracted when the original sample is alkaline and added when acid.

(2) Determine the carbonyl content of the sample by the equation:

$$C = \frac{(B - T) \times N \times 0.1}{W} \pm F$$

where

B is the volume in milliliters of standard acid required for the blank.
T is the corresponding volume for the test sample.
W is the weight of this sample in grams.
C is the carbonyl content of the sample, expressed as centi-equivalents per gram of sample.

(3) If the type of the carbonyl compound is known, multiply the value for C by the molecular weight and divide by the number of carbonyl groups in the molecule to give the percentage by weight of that particular compound. If the type of the carbonyl compound is not known, and no further identification is required, calculate all carbonyl compounds as acetaldehyde. In this case, multiply the value for C by 44.05, and report the product as:

Carbonyl compounds, calculated as acetaldehyde, ... per cent

(4) If required by the specifications, multiply the value for C by 561 and report the product as:

Carbonyl number .... mg KOH/gram

DETERMINATION OF PEROXIDES IN ISOPROPYL ETHER

This method was adapted from a similar method for the determination of peroxides in normal butyl ether. The sample is freed of dissolved air and, under an inert atmosphere, potassium iodide is added, and the liberated iodine titrated with thiosulfate. In this version, a correction is made for the iodate content of the potassium iodide reagent.

This method is limited to the determination of peroxides in isopropyl ether.

Reagents

(1) *Potassium Iodide*—This should be either chemically pure or reagent grade with a low content of iodates. It should be finely ground or powdered and kept in a place protected from direct sunlight.

(2) *Sodium Thiosulfate Solution*—Of the chemically pure pentahydrate, 2.5 g is dissolved in recently boiled and cooled distilled water, and the solution made up to 1 liter. After standing for 2 weeks, the solution is standardized by accepted analytical technique against potassium dichromate or potassium iodate at frequent intervals.

(3) *Glacial Acetic Acid*—Chemically pure or reagent-grade stock is used.

(4) *Sodium Bicarbonate*—The USP grade is satisfactory, but whatever grade is used, the reagent must be finely powdered.

Procedure

(1) Into a 250-ml glass-stoppered Erlenmeyer flask, 25 ml of glacial acetic acid, 5 ml of distilled water, and 25 ml of the ether to be tested are introduced in this order. The flask is swirled to effect mixing. One to two grams of the sodium bicarbonate

is added, and the flask stoppered lightly so that the excess pressure may be easily relieved. No grease or other lubricant should be used on the stopper.

(2) Without delay, 50 ml of glacial acetic acid, 10 ml of distilled water, and 1 to 2 g of bicarbonate are mixed in a 100-ml glass-stoppered graduate and the cylinder lightly stoppered. After 30 to 45 seconds, 2 g of potassium iodide is added and the cylinder shaken vigorously.

(3) One half of the solution obtained under (2) is added to the sample in the flask, and the flask stoppered and shaken. Both the flask and the graduate are then placed in the dark for 10 minutes.

(4) At the end of this time, the solution in the graduate should be colorless. If the solution is colored, it is titrated with the standard thiosulfate solution until colorless, and the volume of thiosulfate recorded. Starch must not be used as an indicator.

(5) In a similar manner, and without the use of starch, the contents of the flask is titrated with the thiosulfate to the disappearance of the brown color of the liberated iodine. The volume is recorded.

Calculations

(1) The volume of thiosulfate used for the blank is subtracted from that used for the sample.

(2) The active oxygen content may then be calculated as follows:

$$\text{(Net volume of thiosulfate)} \times \text{(Normality of thiosulfate)} \times 0.008 \times 4 = \text{grams of active oxygen per 100 ml}$$

This value $\times$ 10,000 = grams of active oxygen per 1,000,000 ml.

L.I.C. No. 300.10B

PEROXIDE NUMBER OF PETROLEUM PRODUCTS

This procedure is suitable for measuring the peroxides which are formed during the storage of petroleum products, particularly cracked gasolines. Two methods will be outlined, both of which were adopted from similar ones appearing in the literature. The first is believed to be more accurate, and the second more rapid.

The method is applicable to clear petroleum products and to most dyed and inhibited oils, since the reaction is observed in the aqueous extract. It may also be used on water solutions containing peroxides if no other substances capable of oxidizing ferrous ions are present.

The PEROXIDE NUMBER is the number of milliequivalents of active oxygen or its equivalent which is present in 1 liter of the sample under test. To express the results in terms of parts per million of active oxygen, multiply the peroxide number by 8.

Both methods are based on the same reaction, which may be expressed as follows:

$$\begin{array}{c} >\mathrm{C{-}C}< \\ |\quad\;| \\ \mathrm{O{-}O} \end{array} + 2FeSO_4 + H_2SO_4 \longrightarrow \begin{array}{c} >\mathrm{C{-}C}< \\ |\quad\;| \\ \mathrm{HO\ OH} \end{array} + Fe_2(SO_4)_3$$

The reaction mixture includes a solution of ammonium thiocyanate, which is colorless in the presence of ferrous, but is red in the presence of ferric, ions.

In the first method, a measured amount of the sample is treated with an excess of the ferrous sulfate solution. If a red color develops, the reaction mixture is titrated with standard titanous chloride solution to the disappearance of the red color. The peroxide number can be calculated from the volume and normality of the titanous solution. In the second method, the sample and the ferrous solution are mixed in a definite ratio, and the intensity of the developed color compared with a set of calibration standards.

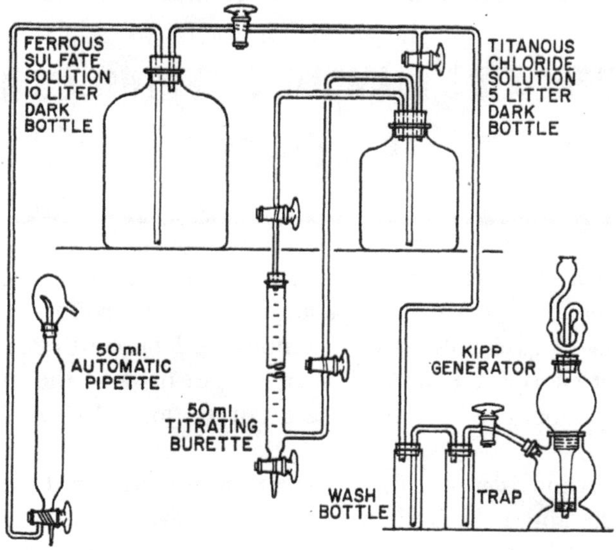

FIGURE 1. Schematic Diagram of Equipment for Peroxide-Number Determination

Apparatus

(1) *Dispenser* as shown in Figure 1. Since the stock solutions are susceptible to air oxidation, both must be kept under a blanket of inert gas. A Kipp generator filled with marble chips and hydrochloric acid will provide a moderate pressure of carbon dioxide at all times. The two smaller bottles may be of any convenient size. The first serves as a spray trap, while the second is filled with copper sulfate to remove any hydrogen sulfide. Any other arrangement may be used to protect both solutions from oxidation.

(2) *Stock Bottles* made of dark glass. The size of the bottles depends on the anticipated demand for the test. Generally, a 10-liter bottle for the ferrous sulfate and a 5-liter one for the titanous chloride are satisfactory.

Reagents

(1) *Indicator*—Dissolve 100 g of chemically pure ammonium thiocyanate in 400 ml of distilled water, and store the solution

in a dark-colored bottle. The bottle should be provided with a cork fitted with a 1-ml pipette having a rubber bulb on the open end.

(2) *Potassium Dichromate*—Dissolve exactly 0.4903 g of chemically pure potassium dichromate in distilled water and dilute to 1 liter. If carefully prepared, this solution will be 0.01000 N, and may be used as a primary standard.

(3) *Ferrous Sulfate Solution*—Into the 10-liter stock bottle, put 5 liters of methanol, 3.5 liters of distilled water, and 5 g of pure iron wire. Purge the bottle of air by bubbling carbon dioxide through a tube reaching almost to the bottom of the bottle. In a separate container, put 2 liters of distilled water and add, slowly and with constant stirring, 50 ml of concentrated sulfuric acid. Purge the flask of air with carbon dioxide. Add 5 g of pure iron wire and 91.5 g of chemically pure ferrous sulfate heptahydrate. Warm until solution is complete, excluding air during this operation. When solution is complete, remove a few milliliters with a pipette and test for the presence of ferric ions with the thiocyanate indicator. When the solution is free of ferric ions, cool it rapidly to room temperature, transfer to the stock bottle, and attach the bottle to the assembly.

(4) *Titanous Chloride Solution*—Slowly and with constant agitation, pour 250 ml of concentrated sulfuric acid into 5 liters of distilled water. After cooling, transfer the solution to the stock bottle and purge with carbon dioxide. Add 30 ml of titanous chloride solution (15 to 16 per cent) and continue the purging to effect complete mixing. Connect the bottle to the assembly as shown in Figure 1.

To standardize, put 50 ml of the stock ferrous solution and 1 ml of the indicator into a 125-ml Erlenmeyer flask. If a red color should appear, cautiously add $TiCl_3$ until the red color is just discharged. Return the level of the titanous solution to the zero mark on the burette, and disregard the amount added. When the solution is colorless, add exactly 25 ml of the 0.01 N dichromate and without delay titrate with the titanous solution to the disappearance of the red color. Note the volume of the titanous solution and calculate its normality by the equation:

$$\frac{\text{Volume of dichromate solution} \times \text{Normality}}{\text{Volume of titanous solution}} = \text{Normality of } TiCl_3 \text{ solution}$$

(5) *Standard Color Samples*—Prepare separately three basic solutions by dissolving the indicated amounts of the reagents in Table 1 in enough hydrochloric acid (3 to 97) to make 1 liter of solution:

TABLE 1

Reagent	Grams
Cobaltous chloride hexahydrate	59.49
Ferric chloride hexahydrate [1]	45.05
Cupric sulfate pentahydrate	62.43

When properly prepared, each of the three basic solutions will be 0.5 normal. Prepare a set of standards by mixing these solutions in the ratios given in Table 2.

TABLE 2

Peroxide Number	$CoCl_2$	$FeCl_3$	$CuSO_4$
0.1	7.0	5.5	0.5
0.2	10.0	6.0	0.5
0.3	18.0	7.5	0.5
0.4	25.0	8.0	0.5
0.5	34.0	8.5	0.5
0.6	40.0	10.0	1.0
0.7	50.0	11.0	1.0
0.8	62.0	12.0	1.0
0.9	75.0	12.0	1.0
1.0	87.0	12.0	1.0

For each standard solution, mix the reagents in the indicated numbers of milliliters and dilute to 100 ml with dilute hydrochloric acid prepared by mixing 3 ml of the concentrated acid with 97 ml of water. Store the solutions in well-stoppered 4-oz

[1] Since ferric chloride crystals are extremely hygroscopic, the exact composition is always uncertain. The stock ferric solution may be prepared by dissolving 0.9313 g of pure iron wire in 6.6 ml of concentrated HCl and 30 ml of distilled water. After solution is complete, add 10 ml of 3 per cent hydrogen peroxide, boil until the odor of chlorine has disappeared, cool, and dilute to exactly 100 ml.

bottles made of clear flint glass. The standards should be stable for at least 2 years.

It is permissible to measure the color of each of the standards in some type of colorimeter and prepare a graph, plotting the per cent transmittance against peroxide number. Obviously, the same procedure should be followed for the examination of the standard and the unknown solutions.

Procedure—A

(1) Measure 50 ml of the ferrous sulfate and 1 ml of the indicator solutions into a 125-ml Erlenmeyer flask. If a red color should develop, cautiously add the $TiCl_3$ solution until the color just disappears. Return the level of the titanous solution to the reference mark on the burette and disregard the amount added.

(2) Without delay, pipette 10 ml of the sample under test into the flask. Insert the stopper and shake the flask for 3 minutes.

(3) If a red color develops, titrate the contents of the flask with the standard titanous solution until the color is discharged, shaking vigorously as the end point is neared. Note and record the necessary volume of the titanous solution.

(4) Calculate the peroxide number by the equation:

$$P = \frac{V \times N \times 1000}{S}$$

where

V is the volume in milliliters of the titanous solution used in the titration.
N is the normality of the titanous solution.
S is the volume of the sample, in milliliters.
P is the peroxide number of the test sample.

(5) If the peroxide number is less than 1, repeat the entire procedure, using 50 ml of the sample.

(6) If the final result, even with the larger sample, is less than 0.1, report the peroxide number as *trace*. In all other cases, report to the nearest 0.1 unit.

The following precautions must be rigidly observed:

(a) The line from the titanous chloride stock bottle must be flushed before using if an hour or more has elapsed since the last test.

(b) The normality of the titanous solution must be checked frequently. This must be done daily when the solution is fresh and later on at weekly intervals.

(c) The titrations of samples must be completed as rapidly as possible, since a longer time will increase the observed values.

(d) If the sample contains dyes or inhibitors, put the mixture obtained under (2) of PROCEDURE into a separatory funnel and allow the layers to separate. Draw off the aqueous layer into the original flask, add 25 ml of benzene, and continue.

Procedure—B

(1) Put into a 4-oz sample bottle 100 ml of the ferrous sulfate and 2 ml of the indicator solutions. If a red color develops, cautiously add the titanous solution until the color is just discharged, avoiding an excess. Disregard the amount added. (The bottle must be of the same type and dimensions as those used for the standard color samples.)

(2) Add 10 ml of the test sample, stopper the bottle, and shake for 2 minutes. Put the bottle on the bench top in an upright position and allow the layers to separate.

(3) Compare the color of the sample with the set of standards and determine the closest match. If the observed value is greater than 1, repeat the test, using a smaller volume of sample.

(4) Repeat the same procedure as a blank on the reagents. Report the difference between the results observed for the sample and the blank as the *peroxide number* of the sample under test.

DETERMINATION OF THE RESIDUAL ODOR OF HEAVY PETROLEUM SOLVENTS

The odor of a petroleum distillate which is to be used as an industrial solvent is sometimes of critical importance. In such cases, it is possible that the product itself might have an odor which is so faint as to be negligible, but, on evaporation, the odor of the residue would be objectionable. In this method, the sample is evaporated to a very low volume and the odor taken on the last few milliliters.

This method is suitable for the examination of heavy solvents, such as V.M. & P. naphtha, mineral spirits, etc.

Apparatus

(1) *Evaporating Dish*—Pyrex glass of about 150-ml capacity. Before use, weigh 5 oz of a heavy solvent into the dish and make a permanent mark, indicating the liquid level on two sides of the dish.

(2) *Perfume Paper* measuring $\frac{1}{4}$ by 6 in., made of absorbent paper. A satisfactory product is the *perfume testing blotter,* which is available from Frank Orlandi, 20 Beekman Street, New York 7, N. Y. If not available, a strip of the proper dimensions may be cut from a sheet of filter paper. It is important that the paper is protected from laboratory odors.

Procedure

(1) Measure about 100 g of the test sample into a clean, dry evaporating dish. Put the dish on a steam bath and leave it there until 95 per cent of the sample has evaporated.

(2) Remove the dish from the bath and without delay dip a strip of paper into the liquid so that about 1 in. of the paper is immersed. Remove the strip and hold it in the air until most of the solvent has evaporated and the line of demarcation be-

tween the wetted and unwetted portions just disappears. Determine the odor of the residue on the paper by smelling. Repeat the dipping and smelling as many times as necessary until the presence or absence of objectionable odors is confirmed.

In making this observation, gentle sniffs are better than deep ones; if the product contains any definite odor, a deep sniff may cause temporary paralysis of the olfactory nerves. If more than one sample is under test, pause a few moments between samples.

(3) Rate the odor of the sample according to Table 1.

TABLE 1

Passes	Practically odorless, or at most faintly aromatic.
Does Not Pass	Odor of garlic, onion, hydrogen sulfide, phenol, kerosene, cracked naphtha, or nonhydrocarbon substances.

AMINO-TYPE INHIBITORS IN GASOLINE

Alkyl-substituted aminophenols and *p*-phenylenediamine compounds are widely used in gasoline to increase storage stability and prevent gum formation. Disecondary butyl *p*-phenylenediamine, which is sold commercially under the trade names of U.O.P. No. 5, DuPont No. 22, and Tenamene 2, is a popular type. This method may be used for the examination of gasolines containing any one of these phenylenediamines. It may also be used for the determination of aminophenols, with proper changes in the standardization.

It should be clearly understood that the procedure described will measure only the *active* inhibitor in a sample. In some cases, particularly when the sample is a thermal-cracked gasoline, the inhibitor may be consumed at a rapid rate. For plant-control purposes, the sample should, therefore, be examined as soon as practicable after the inhibitor has been added.

A portion of the sample is treated with dilute acid to extract the inhibitor. The acid extract is adjusted to a given *p*H value and treated with hydrogen peroxide. This will result in a pink color, the intensity of which is proportional to the concentration of the inhibitor. The intensity is measured on a colorimeter, and the amount of the inhibitor determined by comparison with a calibration curve.

Apparatus

(1) *Spectrophotometer*—Any convenient type may be used on which readings can be obtained in the general region of 500 mμ.

Reagents

(1) *Hydrochloric Acid, 0.1 N*—Dilute 7.5 ml of the concentrated acid to 1 liter with distilled water. The exact strength is not critical, and the solution need not be standardized.

(2) *Hydrogen Peroxide, 30 per cent*—Merck's Superoxol is satisfactory.

(3) *Buffer Solution*—Dissolve 100 g of ammonium acetate in distilled water, add 5 ml of glacial acetic acid, and dilute to 1 liter.

(4) *Ethyl Acetate.*

(5) *Standard Sample* available from the Standard Inspection Supply under SIS Item 4512. The sample consists of about 5 ml of U.O.P. No. 5, stored under nitrogen in a sealed glass ampoule. So far, no discernible difference has been found between U.O.P. No. 5, duPont No. 22, and Tenamene 2. The same standard sample may, therefore, be used for the determination of any one of these inhibitors.

Precautions for Handling the Standard Sample

(1) At all times, protect the standard sample from light. Do not open the ampoule until everything else is ready and complete the standardization as soon as possible.

(2) The liquid sample will cause painful irritations of the skin, and the vapors are injurious to the lungs. When opening the ampoule, wear rubber gloves, a rubber apron, and safety goggles. If the liquid should come into contact with the skin, remove it immediately by washing with isopropyl alcohol followed by a thorough washing with soap and water. Clothing contaminated with the inhibitor should be removed without delay and dry cleaned before being worn again.

Calibration

(1) Fill a 500-ml volumetric flask about two thirds full with benzol. Unpack and open an ampoule of the standard sample and draw about 200 mg of the liquid into a Lunge pipette. (*Caution:* do not fill the pipette by sucking with the mouth. Use a rubber bulb or house vacuum line.) Place some 120.0 mg of the sample into the volumetric flask, the weighings being accurate to the nearest 0.1 mg. Fill the flask to the mark with more benzol and shake it thoroughly.

(2) Using 100-ml volumetric flasks and pipettes or a small burette, dilute the solution further according to Table 1.
Prepare at least five of these dilute solutions. Do not allow any of the solutions to stand for more than 24 hours.

TABLE 1

Milliliters of Solution in Benzol to Make 100 ml	equals	Pounds of Inhibitor per 5000 gal
1.0		0.1
3.0		0.3
5.0		0.5
7.0		0.7
9.0		0.9

Table 1 is based on exactly 120 mg of the standard sample made up to 500 ml with benzol, in which case, each milliliter of the stock solution will contain 0.24 mg of the inhibitor. It is, however, difficult to weigh any predetermined amount from the Lunge pipette. If the weight of the standard sample differs from 120 mg, adjust the value in the last column by the equation:

$$120 : W = X : Y$$

where

W is the weight of standard sample actually taken, in milligrams.

X is the value shown in the last column of Table 1.

Y is the adjusted value, based on the actual weight of sample.

As an example, let us assume that the weight of inhibitor was 130 mg instead of 120, and that 3 ml of the stock solution was diluted to 100 with benzol. The equation would then be:

$$120 : 130 = 0.3 : Y, \text{ or } 0.325 \text{ lb/5000 gal}$$

(3) Transfer the entire contents of one of the flasks to a separatory funnel, add 15 ml of 0.1 N HCl, and shake vigorously for 2 minutes. Draw off the acid layer into a 100-ml volumetric flask, add another 15 ml of the acid, and shake for 1 minute. Draw off this layer into the volumetric flask and add buffer solution until a total volume of about 95 ml is obtained. Add 1 ml of the 30 per cent hydrogen peroxide and fill the flask to the mark with the buffer solution.

(4) Without delay, transfer a portion of the solution to the

cell of the instrument and read the optical density or per cent transmittance following the manufacturer's instructions for the model in use. Keep the solution under constant observation for a few moments and note its maximum color intensity. Record the reading of the instrument and the concentration of the inhibitor in this sample.

If a spectrophotometer is used for this measurement, set the dial at a wave length of 500 mμ. If a filter-type photometer is used, insert the green filter before measuring the color. The color will gradually increase to a maximum and then start to fade. As the color becomes more intense, the per cent transmittance will decrease, and vice versa. The maximum color intensity is, therefore, the time when the transmittance is at the minimum.

(5) Repeat with each of the other standard samples, recording the maximum color intensity of each. Prepare a standard graph, plotting the concentration of the inhibitor against color development.

Procedure

(1) Extract 100 ml of the sample under test with 15 ml of the 0.1 N HCl by shaking vigorously in a separatory funnel for 2 minutes. Allow to settle, draw off the acid layer, and extract the sample again, shaking with 15 ml of acid for 1 minute. Combine the acid washings.

If the amount of sample is small, it is permissible to use proportionately smaller amounts of sample and acid. Bear in mind, however, the dilution ratio of the sample and the final solution.

(2) If the acid extract is colorless, proceed as previously described, beginning with (3) under CALIBRATION. If the extract is colored or cloudy, transfer to a separatory funnel and wash with successive 10-ml portions of ethyl acetate until the solution is clear. Usually one or two washings will suffice. When the solution is clear, treat as previously described.

(3) Determine the inhibitor content of the sample by comparison with the standard curve, and report as:

Inhibitor Content .... lb per 5000 gal
(amino type, as U.O.P. No. 5)

Tetraethyl lead will not interfere, nor will any of the dyes commonly used in gasoline or other organic compounds, such as phenol, cresol, catechol, resorcinol, or thymol.

If preferred, the inhibitor may be extracted with a 5 per cent solution of acetic acid, rather than the 0.1 N hydrochloric acid specified in the text. In some cases, the use of acetic acid rather than hydrochloric may give results which correlate more closely with the *oxidation stability test.*

The type of inhibitor used in the preparation of the calibration curve must be the same as in the test sample. The standard sample sold by the SIS will be of the phenylenediamine type, since this inhibitor is the one most commonly used in the United States. If the sample contains an aminophenol, the same product must be used for calibration and must be procured locally.

DETERMINATION OF *tert*-BUTYLCATECHOL INHIBITOR IN UNSATURATED HEAVY HYDROCARBONS

p-tert-Butylcatechol (TBC) is frequently added to unsaturated hydrocarbons in concentrations below 0.1 per cent as an antioxidant. Although there are several methods available for the determination of this inhibitor (ASTM D-1157 is an example) the existing procedures are all applicable only to the lighter hydrocarbons, such as butadiene and isoprene. This method was, therefore, developed to permit the examination of heavier products. It is applicable to heavy hydrocarbons where the evaporation of the product to isolate the inhibitor presents a problem. Only TBC in an active, unoxidized form appears to be determined by this procedure.

Other phenolic substances present will interfere. In aged samples, some oxygenated substances will accumulate and will be extracted along with the TBC. By study of the optical absorbancy of this yellow color in the absence of TBC, an empirical correction for this interference can be made.

The inhibitor is extracted from the sample by means of an alcoholic solution of sodium hydroxide and oxidized to a red-colored compound, probably either a quinone or a hydroquinone. By measuring the intensity of the red color, the concentration of the original inhibitor can be determined by comparison with a set of calibration standards.

Apparatus

(1) *Colorimeter*—The recommended instrument is the Beckmann spectrophotometer, model DU, equipped with 1-cm cells.

(2) *Weighing Bottles*—Glass stoppered, of about 12-ml capacity. Satisfactory bottles are available from the Fisher Scientific Co. under their catalog number 3-410.

(3) *Centrifuge*—Any convenient type may be used, in which the weighing bottles may be handled without danger of breakage.

(4) *Pipette*—Of 0.2-ml capacity, graduated in 0.01-ml divisions.

Reagents

(1) *Alcoholic Sodium Hydroxide Solution*—Saturate 100 ml of absolute methanol with chemically pure sodium hydroxide pellets at room temperature. The final solution should be about 0.9 N and free of suspended matter and color.

(2) *Petroleum Ether.*

(3) *tert-Butylcatechol* (TBC) manufactured by the Dow Chemical Company.

CAUTION: Some people are allergic to the inhibitor. This product should, therefore, be handled with care, avoiding contact of either the compound or its solutions with the skin. Wash the hands frequently with soap and water.

Calibration

(1) Dissolve 0.05 g of TBC in 5 ml of absolute methanol and make up to 10 ml. Pour 5.0 ml of the alcoholic caustic solution into each of four small glass-stoppered bottles. From a small graduated pipette, add the following volumes of the TBC solution listed in Table 1 to separate bottles:

TABLE 1

Milliliters of TBC Solution	Final TBC Concentration g/liter
0.040	0.040
0.080	0.079
0.120	0.117
0.160	0.155

(2) Stopper the bottles and shake for 2 minutes. Pour a portion of each solution into a 1-cm cell and determine the optical absorbancy at 570 and 625 mμ, using absolute methanol in the

reference cell. Subtract the absorbancy at 625 from that at 570 and plot the net values on a curve.

Procedure

(1) Weigh accurately a 1 to 2-g portion of the sample under test into a small glass-stoppered bottle. Add 2 ml of the alcoholic caustic solution and shake for 2 minutes.

(2) Add 2.0 of absolute methanol, 0.4 ml of distilled water, and sufficient pertoleum ether to cause the hydrocarbon layer to form an upper phase. Centrifuge the mixture until the layers have separated and, by means of a medicine dropper, remove and discard the upper (petroleum ether) layer.

(3) Wash the alcoholic extract twice more with fresh portions of petroleum ether, removing the ether phase after each washing. Note and record the final volume of the alcoholic solution.

The volume may be measured directly, or calculated from the weight and the specific gravity.

(4) Transfer a portion of the alcoholic solution to a 1-cm cell and determine the absorbancy at the same settings as used in the calibration, with absolute methanol in the reference cell. Subtract the absorbancy at 625 from that at 570 and record the net absorbancy of the sample.

If the color is too intense for a direct measurement, dilute the extract with absolute methanol. Note the final volume of the diluted solution and use this volume in the subsequent calculations.

Calculations

Determine the TBC content of the alcoholic extract from the calibration curve. The TBC content of the sample can be determined by equation 1 or 2.

$$P = \frac{C \times V \times 100}{W \times 1000} \quad \ldots\ldots\ldots\ldots 1$$

where

C is the concentration of TBC in the extract, as determined from the curve.

V is the volume in milliliters of the alcoholic extract.

W is the weight of the sample, in grams.

P is the concentration of TBC, expressed as weight per cent.

$$M = \frac{C \times V \times 10^6}{W \times 1000} \qquad \ldots\ldots\ldots\ldots 2$$

where

M is the concentration of TBC, expressed as parts per million.

Duplicate results should not differ from the mean by more than 1 per cent of the mean. Test results should not differ from the theoretical values by more than 10 per cent.

Many unsaturated hydrocarbons have more or less intense absorption in the same general ultraviolet range as the TBC, so that the separation of the inhibitor is essential.

With regard to the use of TBC as an inhibitor, it should be stated that there is considerable evidence that the TBC is consumed by preformed peroxides and also by air storage, unless precautions are taken. The best practice is to add the inhibitor soon after the distillation of a product and to exclude air during storage.

FLASH POINT BY THE TAG OPEN CUP

The method recommended for determining flash points by the tag open cup is exactly the same as that described in the current version of ASTM D 1310.

ASTM DESIGNATION: D 1310–56 T

Scope

1. (*a*) This method describes a test procedure for the determination of open-cup flash points of volatile flammable materials having flash points below 175°F.

(*b*) This method, when applied to paints and resin solutions which tend to skin over or which are very viscous, gives less reproducible results than when applied to solvents.[1]

Outline of Method

2. The sample is placed in the cup of a Tag open tester, and heated at a slow but constant rate. A small test flame is passed at a uniform rate across the cup at specified intervals. The flash point is taken as the lowest temperature at which application of the test flame causes the vapor at the surface of the liquid to flash, that is ignite but not continue to burn.

---

1 A value in excess of 80°F by this method has been established by the U. S. Interstate Commerce Commission as the minimum value for classifying liquids as nonflammable. Materials having a flash point at or below that value must be identified with an ICC Red Label for interstate shipment by common carrier within the USA.

For details of ICC shipping regulations, see *Tariff 8,* issued March 15, 1951, by H. A. Campbell, agent, 30 Vesey St., New York 7, N. Y., publishing Interstate Commerce Commission Regulations for Transportation of Explosives and Other Dangerous Articles by Land and Water in Rail Service and by Motor Vehicle (highway and water) including Specifications for Shipping Containers. (Regulations for transportation in rail express and rail baggage services are also included for information.)

Apparatus

3. (*a*) *Flash Tester*—Tag open-cup tester (Figure 1) as described in detail in Appendix 1.

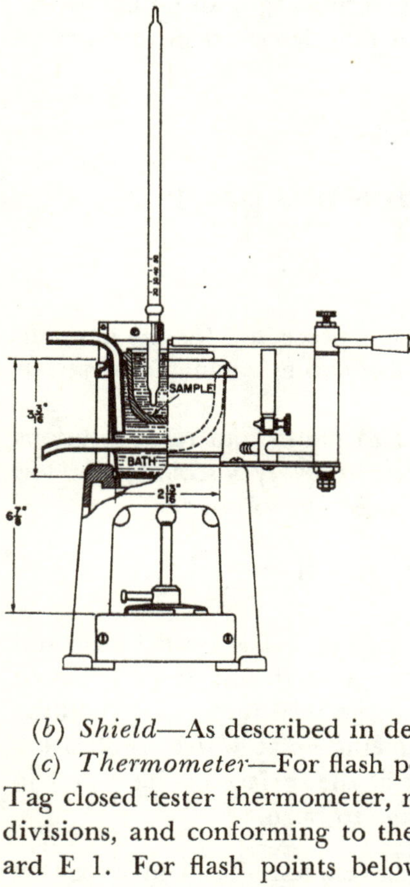

FIGURE 1. Tag Open-Cup Flash Tester

(*b*) *Shield*—As described in detail in Appendix 1.

(*c*) *Thermometer*—For flash points above 40°F, use the ASTM Tag closed tester thermometer, range of +20 to +230°F, in 1°F divisions, and conforming to thermometer 9°F of ASTM Standard E 1. For flash points below 40°F, use ASTM Tag closed tester, low range, thermometer 57°F.[2]

Procedure

4. (*a*) Place the tester on a solid table free of vibration, in a location free of perceptible draft, and in a dim light.

[2] The original Tag open-cup thermometer is a permissible alternate. It is calibrated to −20°F.

(*b*) Run water, brine, or water-glycol solution into the bath to a predetermined level, which will fill the bath to $\frac{1}{8}$ in. below the top when the cup is in place. An overflow is permissible for water-level control.

(*c*) Firmly support the thermometer vertically halfway between the center and edge of the cup on a diameter at right angles to the guide wire, or on a diameter passing through the center of the cup and the pivot of the taper. Place so that the bottom of the bulb is $\frac{1}{4}$ in. from the inner bottom surface of the cup.[3]

(*d*) Fill the glass cup with the sample liquid to a depth just $\frac{1}{8}$ in. below the edge, as determined by the leveling device.

(*e*) Place the guide wire or swivel device in position, and set the draft shield around the tester so that the sides form right angles with each other and the tester is well toward the back of the shield.

(*f*) If a guide wire is used, the taper, when passed, should rest lightly on the wire, with the end of the jet burner just clear of the edge of the guide wire. If the swivel type holder is used, the horizontal and vertical positions of the jet are so adjusted that the jet passes on the circumference of a circle, having a radius of at least 6 in., across the center of the cup at right angles to the diameter passing through the thermometer, and in a plane $\frac{1}{8}$ in. above the upper edge of the cup. The taper should be kept in the "off" position, at one end or the other of the swing, except when the flame is applied.[4]

(*g*) Light the ignition flame and adjust it to form a flame of spherical form matching in size the $\frac{5}{32}$ in. sphere on the apparatus.

(*h*) Place the heater flame under the center of the bath, and adjust it so that the temperature of the sample increases at a rate of 2 ± 0.5°F per minute.[5]

---

3 If the old Tagliabue thermometer is used, immerse to cover well the mercury bulb, but not the wide body of the thermometer.

4 The height of the taper arm is very important. Raising the taper 0.01 in. increases the flash about 2°F. Therefore, if it is suspected that the taper arm has been jarred or bent, the apparatus should be recalibrated. Each unit of apparatus should have its calibration checked about once a week if in constant use, or on each occasion of use if used only occasionally. The calibration procedure provided in this method will cancel out the effect of barometric pressure if calibration and tests are run at same pressure.

5 With viscous materials, this rate of heating cannot always be obtained.

Initial Test

5. Determine an approximate flash point by passing the taper flame across the sample at intervals of 2°F. Each pass must be in one direction only. The time required to pass the ignition flame across the surface of the sample should be 1 second.[6, 7]

Recorded Tests

6. Repeat the procedure by cooling a *fresh* portion of the sample, the glass cup, the bath solution, and the thermometer at least 20°F below the approximate flash point. Resume heating, and pass the taper flame across the sample at two intervals of 5°F and then at intervals of 2°F until the flash point occurs.

Standardization

7. (*a*) Make determinations in triplicate on the flash point of standard paraxylene and of standard isopropyl alcohol which meet the specifications set forth in Appendix 2. Average these values for each compound. If the difference between the values for these two compounds is less than 15°F (8.5°C) or more than 27°F (16°C), repeat the determinations or obtain fresh standards.

(*b*) Calculate a correction factor as follows:

Observed flash of paraxylene = A
Observed flash of isopropyl alcohol = B

$$X = 92 - A$$
$$Y = 71 - B$$

$$\text{Correction} = \frac{X + Y}{2}$$

Apply this correction to all determinations. Half units in correction shall be discarded.

---

[6] Remove bubbles from the surface of the sample liquid before starting a determination. Meticulous attention to all details relating to the taper, size of taper flame, and rate of passing the taper is necessary for good results.

[7] When determining the flash point of viscous liquids and those liquids which tend to form a film of polymer, etc., on the surface, the surface film should be disturbed mechanically each time before the taper flame is passed.

Reporting Data

8. The average of not less than three recorded tests, other than the initial test, shall be reported.

Precision

9. (*a*) For hydrocarbon solvents having flash points between 60° and 110°F, the repeatability is ±2°F and the reproducibility is ±5°F.

(*b*) If results from two laboratories differ by more than 10°F, they shall be considered uncertain and should be checked.

APPENDIX 1

Apparatus

A1. The Tag open-cup tester is illustrated in Figure 1. It consists of the following parts, which must conform to the dimensions shown, and have the additional characteristics as noted:

(*a*) *Copper Bath,* preferably equipped with a constant-level overflow so placed as to maintain the bath liquid level $\frac{1}{8}$ in. below the rim of the glass cup.

(*b*) *Thermometer Holder*—Support firmly with ring stand and clamp.

(*c*) *Glass Test Cup* (Figure 2), of molded clear glass, annealed, heat resistant, and free from surface defects.

(*d*) *Leveling Device,* or guide, for proper adjustment of the liquid level in the cup (Figure 3). This shall be made of No. 18-gage polished aluminum, with a projection for adjusting the liquid level when the sample is added to exactly $\frac{1}{8}$ in. below the level of the edge or rim of the cup.

(*e*) *"Micro," or Small Gas Burner* of suitable dimensions for heating the bath. A screw clamp may be used to help regulate the gas. A small electric heater may be used.

(*f*) *Ignition Taper,* which is a small, straight, blow-pipe type gas burner.

(*g*) Alternate methods for maintaining the ignition taper in a fixed horizontal plane above the liquid may be used, as follows:

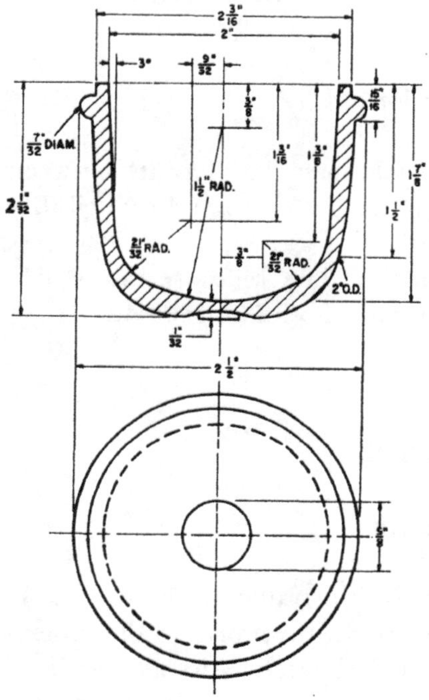

FIGURE 2. Glass Test Cup

(*1*) *Guide Wire,* $\frac{3}{32}$ in. in diameter and $3\frac{1}{2}$ in. in length, with a right-angle bend $\frac{1}{2}$ in. from each end. This wire is placed snugly in holes drilled in the rim of the bath so that the guide wire is $\frac{5}{8}$ in. from the center of the cup and resting on the rim of the cup.[8]

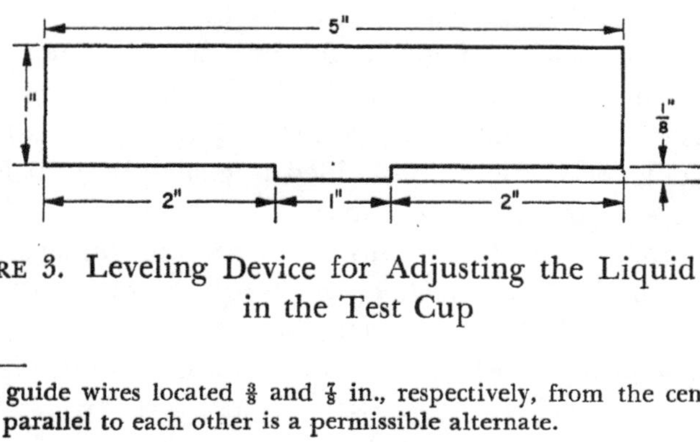

FIGURE 3. Leveling Device for Adjusting the Liquid Level in the Test Cup

[8] Two guide wires located $\frac{5}{8}$ and $\frac{7}{8}$ in., respectively, from the center of the cup and parallel to each other is a permissible alternate.

(2) *Swivel-Type Taper Holder,* such as is used in ASTM Method D 92. The height and position of the taper is fixed by adjusting the holder on a suitable ring-stand support adjacent to the flash cup.

(*h*) *Draft Shield,* consisting of two rectangular sheets of non-combustible material, 24 by 28 in., are fastened together along the 28-in. side, preferably by hinges. A triangular sheet, 24 by 24 by 34 in., is fastened by hinges to one of the lateral sheets (to form a top when the shield is open). The interior of the draft shield shall be painted a flat black.

APPENDIX 2

Specifications for *p*-Xylene and Isopropanol

p-Xylene, Flash-Point Check:

A2. *p*-Xylene shall conform to the following requirements:

Specific Gravity 15.56°/15.56°C. . . . . .0.860 minimum, 0.866 maximum.

Boiling Range. . . . . .2°C maximum from start to dry point. The range shall include the boiling point of pure *p*-xylene, which is 138.35°C (281.03°F).

Purity. . . . . .95 per cent minimum.

Isopropanol, Flash-Point Check Grade:

A3. Isopropanol shall conform to the following requirements:

Specific Gravity. . . . . .0.8175 to 0.8185 at 20°/20°C as determined by means of a calibrated pycnometer.

Distillation Range. . . . . .shall entirely distill within a 1.0°C range which shall include the temperature 80.4°C.

ALPHA-NAPHTHOL IN ETHER

α-Naphthol is often used as an antioxidant in various chemical products. This method is intended to measure the amount of active inhibitor in ethyl and/or isopropyl ether.

A measured volume of the sample is treated with diazotized naphthionic acid under specified conditions of time and temperature in the presence of isopropyl alcohol. As a result, a color complex is formed, and the intensity of the color is proportional to the naphthol concentration. The color intensity is measured and the weight of inhibitor determined from a calibration curve.

Apparatus

(1) *Colorimeter*—Any convenient type may be used on which readings can be made in the general region of 500 mμ.

(2) *Pipettes* of 1, 2, and 5-ml capacity. The 1-ml size should be graduated in 0.1-ml increments.

Reagents

(1) *Solution A*—Weigh 0.0500 ± 0.0001 g of α-naphthol (Eastman White Label grade) into a convenient container. Dissolve in ethyl ether, transfer the solution quantitatively to a 1-liter volumetric flask, fill the flask to the mark with more ether, and mix thoroughly. The solution must be prepared fresh each week and stored in a tightly stoppered amber bottle in a dark place when not in use. The ether should have less than 5 ppm of peroxides.

(2) *Solution B*—Dissolve 0.340 ± 0.001 g of naphthionic acid (Eastman White Label grade) in 8 ml of 2 per cent hydrochloric acid. Transfer the mixture to a 1-liter volumetric flask and fill to the mark with distilled water.

(3) *Solution C*—Dissolve 0.110 ± 0.001 g of sodium nitrite in distilled water, transfer to a 1-liter volumetric flask, and fill to the mark. This solution must be prepared weekly and stored in the dark when not in use.

(4) *Sodium Hydroxide*—2 per cent aqueous solution.
(5) *Isopropyl Alcohol*—91 per cent.
(6) *Ether*—Containing no inhibitor.

Calibration

(1) Before starting the calibration, turn on the current to the colorimeter and allow the instrument to warm up.

(2) Prepare a special diazo reagent by mixing equal volumes of solutions B and C. Allow the mixture to stand in an ice bath for not less than 10 nor more than 60 minutes before using.

(3) Pour 50 ml of 91 per cent isopropyl alcohol into each of five 100-ml volumetric flasks. Reserve one of these as a blank, and add 1 ml of the uninhibited ether. Using the others, prepare a set of standards containing α-naphthol in the range of 0.00 to 0.10 mg (1 ml of solution A contains 0.05 mg). Pour into each flask 5 ml of the 2 per cent NaOH and mix thoroughly.

(4) Pipette 4 ml of the chilled freshly prepared diazo reagent into each flask and mix thoroughly. Fill each flask to the mark with distilled water and mix again. Allow each solution to stand for 30 ± 1 minutes. This standing time is critical, since the color continues to develop on further standing.

(5) Examine each of the solutions in the colorimeter and prepare a graph, plotting net absorbancy against milligrams of α-naphthol.

If a spectrophotometer is used, find the exact location of the peak height at or near 500 mμ and use this setting in all future work. The curve is conveniently prepared on Codex No. 318 graph paper.

Procedure

(1) Measure 50 ml of 91 per cent isopropyl alcohol into each of three 100-ml volumetric flasks. Reserve one of these as a blank, and add 1 ml of the uninhibited ether. Use the second as a standard and add 1 ml of solution A. Pipette 1 ml of the test sample into the third. Pour into each flask 5 ml of the 2 per cent NaOH solution and mix thoroughly.

(2) Examine each as described under calibration, beginning with the procedure under (4). Subtract the reading of the blank

from each of the others and determine the net naphthol content from the graph.

The standard sample is used as a check on operation conditions, and the observed content should be within 0.002 mg of the calculated content. If the content of the sample under test is beyond the range of the chart, repeat the test using a smaller portion.

(3) Determine the specific gravity of the test sample at 20°C by any convenient means.

Calculation

Calculate the α-naphthol content of the sample by the equation:

$$A = \frac{M/V \times 1000}{S}$$

where

M is the number of milligrams as obtained from the graph.
V is the volume in milliliters of the test portion (normally 1).
S is the specific gravity of the sample.
A is the active α-naphthol content, expressed as parts per million.

Precision

Duplicate results should not differ from each other by more than 2 ppm. The accuracy at the 90 per cent level is of the same order of magnitude.

If this degree of precision is to be obtained, the volume of sample must be measured accurately, and the sample must not be unnecessarily exposed to air or sunlight.

SAMPLING AND TESTING LACQUER SOLVENTS AND DILUENTS

Sampling

From Loaded Tank Cars or Other Large Vessels—The composite sample should be not less than 2 qt and should consist of small samples of not more than 1 qt each, taken from near the top and the bottom by means of a metal or glass container with removable stopper or top. This device, attached to a suitable pole, is lowered to the desired level, when the stopper or top is removed and the container allowed to fill.

From Barrels or Drums—At least 5 per cent of the packages in any shipment should be represented in the sample. The purchaser may increase the percentage of packages to be sampled at his discretion, and it is recommended that every package is sampled in the case of expensive solvents that are bought in small quantities. A portion is withdrawn from the center of each package by means of a "thief" or other sampling device. The composite sample thus obtained should be not less than 1 qt and should consist of equal portions of not less than ½ pt from each package sampled.

Specific Gravity

Any convenient method that is accurate to the third decimal place may be used. The temperature of the sample and the water should be 20°C.

The usual API hydrometers should not be used, since these are calibrated at 60°/60°F. Gravity may be determined by a calibrated specific-gravity hydrometer, or by the procedure under L.I.C. No. 20.02B.

Nonvolatile Matter

100 ml of the sample is placed in a tared porcelain evaporating dish and evaporated almost to dryness on a steam bath. The dish is then heated to constant weight in an oven maintained at 105° ± 5°C. The increase in weight of the dish is the nonvolatile matter and should be expressed in grams per 100 ml.

Color

Refer to L.I.C. No. *40.11A*.

Residual Odor

Refer to L.I.C. No. *310.20A*.

Water by Turbidity

Place 5 ml of the sample in a clean dry 100-ml glass-stoppered graduate. Add, in 5-ml portions, heptane (10° quality, containing not over 5 per cent aromatics) until the total volume specified is obtained, shaking well after each addition and examining for turbidity. Both the sample and the heptane should be at a temperature of 20°C. Water is indicated by turbidity. If turbidity develops, the reference sample should be tested in the same way and compared with the sample.

Water by the Iodine Reagent Method

Refer to L.I.C. No. *150.04A*.

Acidity

Solvents—Using a pipette, 50 ml of the sample is transferred to a small Erlenmeyer flask and titrated with 0.1 N KOH in 99 per cent methyl alcohol, using phenolphthalein as the indicator. The result is reported as milligrams of KOH per gram of sample, the weight of 100 ml of the sample being calculated from the specific gravity.

In the case of esters, the results are also reported as a percentage by weight of the appropriate acid.

Tricresyl Phosphate—100 g of the sample is transferred to a 500-ml Erlenmeyer flask, and 100 ml of neutral alcohol (either ethyl or methyl) added. The mixture is titrated to a phenolphthalein end point, using 0.1 N KOH in methyl alcohol. The percentage of free acid as H_3PO_4 may be found by the equation:

$$A = V \times N \times 0.0327$$

where

V is the volume in milliliters of the standard KOH.
N is the normality of this solution.
A is the free acidity, expressed as percentage of phosphoric acid.

Alkalinity

If the sample is alkaline as received, 50 ml, are pipetted into a small Erlenmeyer flask and titrated with 0.1 N sulfuric acid, using methyl orange as the indicator. The weight of the 50-ml portion is determined from the specific gravity, and the results reported as milligrams of KOH per gram of sample.

In the case of acetone, 100 ml of the sample is pipetted into a small flask, and 0.3 ml of 0.01 N sulfuric acid and 0.5 ml of 0.3 per cent aqueous *p*-nitrophenol added. The resulting solution should not be alkaline to *p*-nitrophenol.

Ester Value

Refer to L.I.C. No. *250.01*. The result obtained should be corrected for any free acidity or alkalinity and then calculated to percentage by weight of the appropriate ester.

Mercury Corrosion Test

Shake 10 ml of the sample with 1 ml of clean mercury in a stoppered test tube and allow to stand for 15 minutes. The mercury should show no discoloration when compared with a blank run with distilled water.

Sulfur

A qualitative test for the presence of hydrogen sulfide and sulfur dioxide is made at the time of performing the distillation test by hanging two strips of paper on the end of the distillation tube. One of these is moistened lead acetate paper, and the other moistened starch iodate paper. If the lead acetate paper is discolored, hydrogen sulfide is present, but no sulfur dioxide. If the lead acetate paper shows no discoloration, but the starch iodate paper develops a blue color, sulfur dioxide is present but no hydrogen sulfide. If both papers are discolored, both gases are present. If neither paper shows discoloration, neither of the gases is present.

Volatile Matter in Tricresyl Phosphate

Weigh 25 g of the sample into a tared flat Petri dish and heat for 3 hours in an oven at 100° to 105°C. The loss in weight represents the volatile matter and should be expressed as a percentage of the original sample.

Flash Point

Refer to L.I.C. No. *10.30E.*

Other Tests

The *Copper Corosion Test; Free Phenols in Tricresyl Phosphate; Toluene Dilution Ratio;* and *Potassium Permanganate Test* are reprinted here from ASTM D-268.

Copper Corrosion Test

Apparatus

The apparatus shall consist of the following:

(*a*) *Flask*—A 300-ml flask of chemically-resistant glass with flat bottom and vial mouth.

(*b*) *Glass Condenser*—A 12-in. glass condenser with the diameter of the condenser tube not less than $\frac{3}{8}$ in. A cork stopper to connect the flask with the condenser.

(*c*) *Copper Strip*—A copper strip $2\frac{1}{2}$ by $\frac{1}{4}$ in. made from No. 20 B. & S. gage, cold-rolled, polished copper, aproximate thickness 0.032 in. and having a $\frac{1}{16}$-in. hole, $\frac{1}{8}$ in. from one end. A piece of soft copper wire about 6 in. in length.

(*d*) *Water Bath*—A water bath of convenient size and design.

PROCEDURE

Polish both sides of the entire copper strip, edges excepted, with a No. 00 emery cloth immediately before using, to avoid discoloration by the atmosphere. Fasten a 6-in. length of soft copper wire through the hole provided near one end of the strip, taking care not to touch the strip with the fingers after polishing. Place the strip in the flask and add 200 ml of the material to be tested. The material must not contain separated water; filter through a dry filter paper, if necessary, to remove water. Connect the flask to the vertical reflux condenser by means of a properly bored cork stopper. It is absolutely necessary that a cork, not rubber, stopper is used, in order to avoid contamination of the material by sulfur from rubber stoppers. The copper wire may be allowed to extend into the condenser tube for convenience in removing the strip. The strip shall be completely immersed and should preferably lie flat and touch the flask only at the ends of the strip. Place the flask in the water bath, which should be just gently boiling, and immerse the flask to the liquid line of the material within the flask. Remove the copper strip 30 minutes from the time the flask was immersed in the bath. Do not touch the copper strip, but remove it by the wire which has been provided. Do not allow the strip to come in contact with separated water during any part of the test, since water causes bad local staining of the copper. If it is desired to preserve the strip for future reference, dip it immediately in white shellac or lacquer. The sheared edges of the test strip shall not be used as a basis for judgment; flat surfaces only shall be considered.

Free Phenols in Tricresyl Phosphate

REAGENTS

(*a*) *Sodium Hydroxide Solution* (*2.5 per cent*).

(*b*) *Sodium Bromide-Bromate Solution* containing 10.09 g of sodium bromide and 2.95 g of sodium bromate per liter.

(*c*) *Potassium Iodide Solution (6 per cent).*
(*d*) *Hydrochloric Acid (sp. gr. 1.18).*
(*e*) *Starch Solution (1 per cent).*
(*f*) *Sodium Thiosulfate (0.0394 N).*

Procedure

Weigh a 10-g sample of the tricresyl phosphate and transfer to a 250-ml volumetric flask; add 50 ml of NaOH (2.5 per cent) at 65°C and shake the flask vigorously for 3 minutes. Dilute this solution to the 250-ml mark and filter. Pipet a 100-ml portion of the filtrate into a 300-ml flask for titration. To the flask add from a buret 15 ml of the sodium bromide-bromate solution and then 5 ml of HCl (sp. gr. 1.18). Add the HCl quickly and put the stopper in place immediately. Shake the flask and allow to stand for 15 minutes. Add 10 ml of the potassium iodide solution and titrate the solution with the 0.0394 *N* sodium thiosulfate solution, using the starch solution as an indicator. Add the starch solution near the end of the titration. The fading out of the blue color marks the end point of the titration. The reactions are as follows:

$$5NaBr + NaBrO_3 + 6HCl = 6NaCl + 3H_2O + 3Br_2$$

$$CH_3C_6H_4OH + 3Br_2 = CH_3C_6HBr_3OH + 3HBr$$

Run a blank titration as described above, using the same reagents.

Calculation

Calculate the percentage of free phenols (cresol) as follows:

$$C = \frac{(a - b) \times 0.00071 \times 100 \times 2.5}{W}$$

where

C = free phenols (cresol) per cent.
a = milliliters of $Na_2S_2O_3$ for titration of the blank.
b = milliliters of $Na_2S_2O_3$ for titration of the sample.
W = weight of sample (10 g).

Toluene (Toluol) Dilution Ratio

Definition

Toulene (Toluol) Dilution Ratio of a solvent is the maximum number of unit volumes of toluene that can be added to a unit volume of the solvent to just cause persistent heterogeneity at a concentration of 8 g of nitrocellulose per 100 ml of combined solvent and toluene.

Reagents

(*a*) *Toluene* (*Toluol*) conforming to the standard specifications for industrial grade toulene for use in paint, varnish, lacquer, and related products.

(*b*) *Nitrocellulose* conforming to the tentative specifications and tests for Soluble Cellulose Nitrate of the American Society for Testing Materials, and in addition of such quality that, when used in determining the toluene dilution ratios of normal butyl acetate and methyl normal propyl ketone, it will give results between the following limits:

	Toluene Dilution Ratio
Normal butyl acetate	2.73 to 2.83
Methyl normal propyl ketone [1]	3.80 to 3.90

Important Factors in Conducting Test

(*a*) *Concentration of Nitrocellulose at End Point*—The dilution ratio increases as the solids content of nitrocellulose at the end point becomes less. It is, therefore, necessary to set an arbitrary solids content as part of the definition of dilution ratio, and for this purpose a solids content of 8 g per 100 ml of solvent plus toluene has been adopted.

(*b*) *Character of End Point*—The end point shall be taken as the first evidence of heterogeneity in the solution. In the case of a gel end point, a time period of 2 minutes shall be allowed for the gel structure to form.

---

[1] The methyl normal propyl ketone shall have a degree of purity of over 99 per cent as determined by the hydroxylamine method.

(*c*) *Temperature of Test*—The test shall be carried out at a temperature of 23° ± 1.1°C (73.4° ± 2°F).

PROCEDURE

(*a*) *Drying of Nitrocellulose*—Every precaution shall be exercised in the drying of nitrocellulose which shall be dried by blowing with hot air at a temperature under 75°C until constant weight within 1 per cent is obtained. Care shall be exercised that the temperature does not exceed 75°C. It is preferable to keep the air at approximately 65°C in order to be safe.

The material shall be removed from the drying apparatus immediately after blowing as an explosion may occur when the material is not removed promptly after the air has been turned off. Glass shall not be used for the drying tube because of the danger of building up a static charge.

(*b*) *Preparation of Solution*—The probable toluene dilution ratio of the solvent shall first be estimated in order to make up the initial nitrocellulose solution in the proper concentration.[2] This may be done by referring to published data on similar materials or by a preliminary determination. A solution shall then be prepared using 5 ± 0.01 g of dry nitrocellulose in the amount of solvent as prescribed in Table 1.

The nitrocellulose shall be weighed into a 125-ml cork-stoppered, tared, Erlenmeyer flask, or other suitable container, on an analytical balance, and then the volume of solvent indicated in the table shall be added from a burette to the nitrocellulose in the flask. Care shall be taken to prevent loss of solvent by evaporation. The stoppered flask shall be shaken or allowed to stand until the nirtocellulose is completely dissolved before starting titration with the toluene. Care shall be taken not to wet the stopper with the solution during the early stages of the titration. When a high concentration of nitrocellulose in solvent is required, dispersion may be more quickly accomplished by adding a portion of the toluene to the solvent, which reduces the solids concentration and thus lowers the viscosity of the solution and increases the rate of dispersion.

---

[2] This procedure for preparation of the solution is specified in order to avoid running a large number of titrations to arrive at a solids content of 8 g of nitrocellulose per 100 ml of combined solvent and toluene at the end point.

(*c*) *Titration*—The toluene maintained at 23° ± 1.1°C (73.4° ± 2°F) shall be titrated into the flask in small additions, the increments being not over 0.5 ml as the end point is approached. After each addition, the flask shall be stoppered and vigorously shaken to disperse any gel or precipitate thrown down by local overconcentration of toluene. The end of the titration shall be considered to have been reached when the first definite evidence of heterogeneity persists at 23° ± 1.1°C (73.4° ± 2°F). With solvents that give a gel end point, a period of 2 minutes shall elapse between the additions of toluene as the end point is approached.

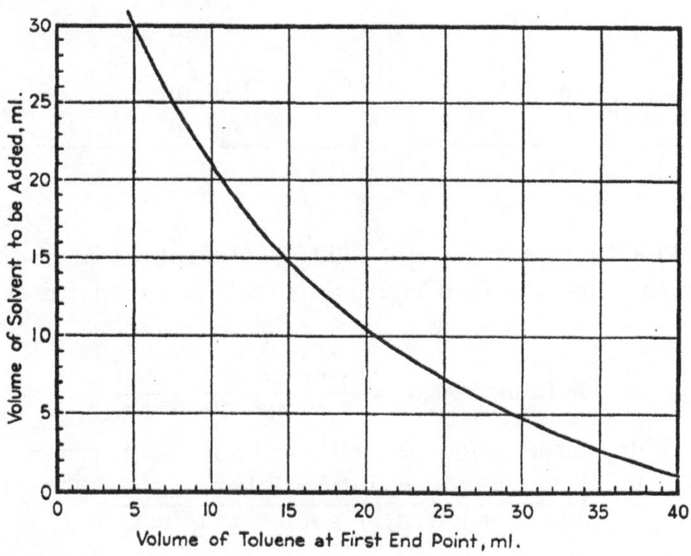

FIGURE 1. Volume of Solvent to Be Added to Complete Titration *versus* Volume of Toluene Used to Reach First End Point

The initial concentration determined from Table 1 has been selected to give approximately 10 g of nitrocellulose per 100 ml of volatiles at the first end point. An additional quantity of solvent shall now be added and the titration continued to the final end point, which shall be reached at a solids concentration of 8 ± 1 g per 100 ml of volatile matter. The correct amount of solvent to add to accomplish this depends on the amount of toluene that was required in the preliminary titration. The curve of Figure 1 is so constructed as to make the volume of solvent to be added direct reading when the volume of toluene used to reach the first end point is known. After the addition of the volume of solvent read from the curve in Figure 1, the titration shall be completed to the final end point.

TABLE 1

Probable Toluene Dilution Ratio	Volume of Solvent per 5 g of Nitrocellulose at 23° ± 1.1°C
1	25.0 ml
2	16.7 ml
3	12.5 ml
4	10.0 ml
5	8.3 ml

CALCULATION [3]

(*a*) The dilution ratios and solids concentrations at both the preliminary and the final end points shall be calculated as follows:

$$\text{Dilution ratio} = \frac{\text{volume of toluene}}{\text{volume of solvent}}$$

$$\text{Solids concentration per 100 ml of volatile matter} = \frac{\text{weight of nitrocellulose}}{\text{volume of solvent} + \text{volume of toluene}}$$

(*b*) A graph shall now be constructed having dilution ratios for the ordinate and solids concentration for the abscissa, and the two sets of values calculated as above shall be plotted as two points on the graph. One of the points shall be very close to 8 g per 100 ml of volatile matter. The correct value at exactly 8 g per 100 ml of volatile matter may be interpolated by drawing a straight line connecting the two experimentally determined points. While the curve showing relationship between dilution

[3] Due to the influence of possible variations in the composition of commercial solvents on the resulting toluene dilution ratio, it is recommended that if data obtained by this method is to be published, the composition of the material under test be stated and the character of the end point be indicated.

ratio and solids concentration is not necessarily a straight line, the error made by interpolating or extrapolating from the straight line connecting the two points is negligible in the proximity of 8 ± 1 g per 100 ml of volatile matter.

INDEX

B

C

D

E

www.ingramcontent.com/pod-product-compliance
Lightning Source LLC
LaVergne TN
LVHW101456200926
848483LV00001B/143